Aerospace Actuators 2

Series Editor
Jean-Paul Bourrières

Aerospace Actuators 2

Signal-by-Wire and Power-by-Wire

Jean-Charles Maré

WILEY

First published 2017 in Great Britain and the United States by ISTE Ltd and John Wiley & Sons, Inc.

Apart from any fair dealing for the purposes of research or private study, or criticism or review, as permitted under the Copyright, Designs and Patents Act 1988, this publication may only be reproduced, stored or transmitted, in any form or by any means, with the prior permission in writing of the publishers, or in the case of reprographic reproduction in accordance with the terms and licenses issued by the CLA. Enquiries concerning reproduction outside these terms should be sent to the publishers at the undermentioned address:

ISTE Ltd
27-37 St George's Road
London SW19 4EU
UK

www.iste.co.uk

John Wiley & Sons, Inc.
111 River Street
Hoboken, NJ 07030
USA

www.wiley.com

© ISTE Ltd 2017

The rights of Jean-Charles Maré to be identified as the author of this work have been asserted by him in accordance with the Copyright, Designs and Patents Act 1988.

Library of Congress Control Number: 2017930553

British Library Cataloguing-in-Publication Data
A CIP record for this book is available from the British Library
ISBN 978-1-84821-942-7

Contents

Introduction ix

**Chapter 1. Electrically Signaled
Actuators (Signal-by-Wire)** 1

    1.1. Evolution towards SbW through the
example of the flight controls............................... 2
        1.1.1. Military applications 2
        1.1.2. Commercial aircraft 3
        1.1.3. Helicopters and compound helicopters 5
    1.2. Incremental evolution from all mechanical
to all electrical.. 9
        1.2.1. Exclusively mechanical signaling 9
        1.2.2. Fly-by-Wire 18
    1.3. Challenges associated with electrical signaling........ 22
        1.3.1. Electrical interfaces.............................. 22
        1.3.2. Evolution of the control and information
transmission architectures................................. 30
        1.3.3. Reliability and backup channels 32
    1.4. The example of landing gears 35

**Chapter 2. Signal-by-Wire Architectures
and Communication** 39

    2.1. Architectures.. 40
        2.1.1. Federated architectures 40
        2.1.2. Integrated modular architectures 41

2.2. Data transmission 43
   2.2.1. CAN 45
   2.2.2. RS422 and RS485 46
   2.2.3. ARINC 429 46
   2.2.4. MIL-STD-1553B 48
   2.2.5. ARINC 629 48
   2.2.6. AS-5643/IEEE-1394b 49
   2.2.7. AFDX (ARINC 664 Part 7) 50
   2.2.8. Triggered time protocol (TTP/C) 52
2.3. Evolutions in data transmission 53
   2.3.1. Power over data and power
   line communication 54
   2.3.2. Optical data transmission
   (Signal-by-Light or SbL) 55
   2.3.3. Wireless data transmission
   (Signal-by-WireLess or SbWL) 58

Chapter 3. Power-by-Wire 59

3.1. Disadvantages of hydraulic
power transmission 60
   3.1.1. Power capacity of hydraulic pumps 61
   3.1.2. Hydraulic pump efficiency 61
   3.1.3. Centralized power generation 62
   3.1.4. Power transmission by mass transfer 62
   3.1.5. Control of power by energy dissipation 63
3.2. Electrical power versus hydraulic power 64
3.3. Improving hydraulically supplied solutions 68
   3.3.1. Reduction of energy losses in actuators 68
   3.3.2. Increased network power density 70
   3.3.3. Other concepts 70
3.4. Concepts combining hydraulics and electrics 71
   3.4.1. Local electro-hydraulic generation 71
   3.4.2. Electro-hydrostatic actuators 73
3.5. All electric actuation (hydraulic-less) 81
   3.5.1. Principle of the electro-mechanical actuator 81

Chapter 4. Electric Power Transmission
and Control 83

4.1. Electric power transportation to PbW actuators 83
   4.1.1. Form 84
   4.1.2. Voltage and current levels 85

4.2. Electric motors	91
4.2.1. Elementary electric machines	91
4.2.2. Conversion of electrical power into mechanical power	95
4.3. Power conversion, control and management	98
4.3.1. Electric power system of a PbW actuator	98
4.3.2. Principle and interest of static switches	100
4.3.3. Groups of switches: commutation cell, chopper and inverter	103
4.3.4. Inverter command	105
4.3.5. The power architecture of a PbW actuator	113
4.4. Induced, undergone or exploited effects	115
4.4.1. Dynamics in presence	115
4.4.2. Torque ripple	118
4.4.3. Energy losses	119
4.4.4. Impact of concepts and architectures on performances	124
4.4.5. Reliability	127
4.5. Integration	130
4.5.1. Overall integration of the actuator	130
4.5.2. Cooling	133
4.5.3. Mechanical architecture of motor control/power electronic units	135

Chapter 5. Electro-hydrostatic Actuators 139

5.1. Historical background and maturing of EHAs	139
5.1.1. PbW actuators with variable displacement pump (EHA-VD)	139
5.1.2. Fixed displacement and variable speed EHA actuators	145
5.2. EHA in service and feedback	159
5.3. EHA specificities	161
5.3.1. Pumps	161
5.3.2. Filling and charging	163
5.3.3. Dynamic increase of mean pressure (pump-up)	164
5.3.4. Energy losses and thermal equilibrium	164
5.3.5. Dissymmetry	168
5.3.6. Control	169

Chapter 6. Electro-mechanical Actuators. 171

6.1. Development and operation of
electromechanical actuators 172
    6.1.1. Space launchers 173
    6.1.2. Flight controls 179
    6.1.3. Landing gears 185
    6.1.4. Helicopters. 191
    6.1.5. Application to engines 194
6.2. Specificities of EMAs. 195
    6.2.1. Power architectures. 196
    6.2.2. Power management functions. 203
    6.2.3. Jamming 206
    6.2.4. Breakage 212
    6.2.5. Thermal equilibrium 214
    6.2.6. Control 214
    6.2.7. Further considerations 217

Bibliography. 219

Notations and Acronyms 235

Index 245

Introduction

This book is the second volume in a series dedicated to aircraft actuators. The first volume, *Aerospace Actuators 1*, focuses on the actuation needs in the aerospace industry, specifically on the reliability and on the hydraulically supplied actuators. This second book, *Aerospace Actuators 2*, is the logical continuation of this. It is, in effect, about the evolution of aircraft towards more (or total) electric systems, and involves "signal" as well as "power". The third volume of the series, *Aerospace Actuators 3*, is dedicated to the detailed analysis of recent achievements. It builds on the concepts and generic solutions that are presented in the first two volumes.

The first two chapters of this book relate to the processing and transmission of signals in electrical form, which will be designated by the general name Signal-by-Wire[1]: the actuator receives and transmits signals electrically. Chapters 3 and 4 relate to the electrical component of the actuators driven by electric power, and categorized by the designation Power-by-Wire. Chapter 5 is dedicated to Electro Hydrostatic Actuators (EHA) and Chapter 6 to Electro-Mechanical Actuators (EMA).

---

[1] The term Signal-by-Wire is used because it is explicit: as it refers to signals transmitted through electrical wire networks. As opposed to the usual name Fly-by-Wire for electric flight controls, which is ambiguous because it does not explain whether it concerns the signals or the power. It was originally introduced to indicate that the flight control actuators receive orders in electrical form and not mechanical. It is often wrongly interpreted as removing the hydraulic power supply component for an electrical power supply. This lack of clarity becomes crucial when including the electrical braking (brake-by-wire) or the steering of the landing gear (steer-by-wire). In the rest of the book, the name Fly-by-Wire therefore exclusively concerns the transmission of orders in electric form to the flight control actuators.

Like the first, this second volume focuses on the needs, the architectures (functional, conceptual and technological), the benefits and the limitations of technological solutions and the orders of magnitude. It uses mathematical models and goes into detail on technological implementation only when it is necessary for the understanding of those principles. This approach allows us to concentrate on the analysis, or synthesis, of technological concepts and their implementation. That being said, it is evident that the capacity for mathematical modeling and a detailed knowledge behind the technology, and the inevitable imperfections of said technology, are major factors in the success of developing and operating industrial products such as actuators. As always, the architectural choices and their implementation are strongly impacted by the capability to make realistic models and by the constraints induced by technology, thereby producing a looped feedback effect on the design process itself.

Throughout this book we will note that the complexity for engineers is greatly increased in the transition to more or total electric actuation, because this feedback loop effect acts in addition to the numerous and new coupled effects that occur between generic topics: mechanical (vibration, tribology, thermal), signal processing (networks, interfaces, control), electric (power electronics, electromagnetic) and other (dependability, human–machine interface), etc. This is why it is particularly important to effectively combine a top-down approach (from the needs to architectures, and then to technological solutions) and a bottom-up approach (from mature technology to architectures). This leads to a hybrid approach, or middle-out. The vision of "requirements, architectures and concepts", is therefore complementary to the vision of the expert to build on a cross-functional approach, with a real system vision in mind. As it turns out, the vision of the expert is well documented in the scientific literature, in contrast to the visions of the designer or system supplier. This book seeks to capitalize and document this system vision applied to more or total electric actuation in aerospace. As such, focused experts will not find a high level of detail in their specialization (signal transmission, power electronics, electric machines). On the contrary, a particular effort will be to focus on popularization, to help the reader to become accustomed to conventional solutions and realize the principles and characteristics of more or total electric actuation. As with Volume 1, the following literature is recommended to accompany this comprehension:

– for more electric actuation in aerospace [RAY 93, SCH 98];

– for electrical systems of aircraft and avionics [COL 11, CRA 08, DAN 15, DUB 13, MOI 08, MOI 13, SPI 14, USF 12, WIL 08, WIL 09];

– for power electronics and electrical machines [DED 11, GIE 10, GRE 97, LAC 99, RAS 11].

NOTES:

– If we adopt a needs/solutions vision, it is interesting that more or full electrics is often presented as an objective[2] (a need) when in fact we should see it as a means (an evolution) to increase performance, reduce constraints and create entirely new services. In commercial aerospace, the final needs of the passenger or even the airline, are neatly summarized by four qualifiers: cheaper, safer, greener and faster. In recent times, this last qualifier tends to be forgotten, because in light of the present technology, an increase in speed impacts negatively on the other qualifiers.

– Although it seems inconsistent, with respect to the advancement towards more or total electric actuation, there is a great need for more research to be done in all fields of mechanical engineering (e.g. solid mechanics, materials, vibrations, tribology, thermal). And this will be a recurrent theme throughout this book.

I.1. Requirements in terms of the actuation for piloting an aircraft

The piloting of an aircraft, initially purely manual, developed actuation functions at both power and signal[3] levels. This was due to some very diverse needs, listed in the following sections, and which are closely related to the aircraft type, mission and the inherent nature of the actuation itself (to do with the flight control, landing gear or engine).

I.1.1. *Reducing the control forces needed for piloting*

It is necessary to limit the forces to be exerted by the pilot, and make them compatible with human capabilities with a view to reduce physical fatigue. This is to ensure that the pilot can apply the necessary level of effort required in the worst piloting situations (during transient) or for longer

---

2 This can be seen as in terms of market positioning.
3 The reader will find further detail on actuation needs in Chapter 1 of Volume 1.

periods (steady state conditions). Table I.1 gives some examples for the maximum levels of control forces conceivable, as defined by European standards. For a large aircraft, note the factor 7.5 to 10 between the short-term and long-term control forces. The specification of the maximum control forces needed to pilot military aircraft [MIL 80, MIL 97] is much more complex because they depend on many factors (phase of the mission, flight quality, load factor, etc.). Reducing control force mainly concerns the power aspect of actuation, which will be discussed in Chapter 3.

Standard	Force applied to the control wheel or pedals (N)	
CS-25.143d Large aeroplanes [EUR 15]	Pitch/Roll	Yaw
Transient, one handed	222/111	667
Transient, both hands	334/222	667
Long duration	44.5/22	89
CS-29.397 Large rotorcraft [EUR 12]	Longitudinal/Lateral	Yaw
	445/298	578

Table I.1. *Examples for specifying maximum control force*

I.1.2. *Reducing the intellectual burden on the pilot*

In order for the pilot to concentrate on the mission at hand, it is important to reduce the intellectual exertion associated with the conduct of the flight and the aircraft. This requirement therefore concerns the development of control commands:

– to stabilize the aircraft (Control Augmentation System or CAS) by rejecting the effects of various disturbances (e.g. gusts or crosswinds);

– able to *decouple, synchronize or coordinate* the different commands to act only on the desired degrees of freedom (e.g. remove the yaw which is induced by the aileron deflection; or further increase the engine throttle of a helicopter when there is an increase in the cyclic pitch);

– to *compensate* the controls to ensure the balance of the aircraft (e.g. action on the angle of attack of the horizontal stabilizer to ensure the longitudinal balance through the pitch trim);

– to *remain within the flight envelope* by monitoring margins compared to permissible limits (e.g. in terms of the load or the never-exceed speed).

I.1.3. *Allowing independent and automatic piloting*

Autopilot (AP) or an Automatic Flight Control System (AFCS) eliminates the need for a human pilot onboard service in conjunction with an autopilot[4]. This occurs:

– either to unburden long and repetitive tasks (e.g. the automatic pilot of an airliner);

– or simply to allow unmanned flight. This therefore applies to both space launch vehicles and missiles, Unmanned Aerial Vehicle (UAV), Optionally Piloted Vehicle (OPV) or Remotely Piloted Aircraft (RPA).

I.1.4. *Increasing the performance of the aircraft*

With human piloting, the issuing of actuator control orders is limited by the accuracy and subjectivity of a pilot's senses, his capacity for processing information in real time (especially under heavy loads) and finally by his ability to react.

This latter limitation can be illustrated by studying the transfer functions of the pilot, that is to say, the mathematical model representing the transfer between perception and action [MIL 97, MC 74, ROS 03]. However, as the pilot "adapts" to the nature (to the transfer function) of the system under control, it is impossible to uniquely define his transfer function. The pilot can for example "correct" the dynamics of the controlled system by introducing a proportional, proportional-integral or lead-lag action. In any case, it is noted that the command applied by the pilot is tainted with a pure delay varying from 0.1 to 0.4 s typically. It is thus clear[5] that it is impossible, for example, to dose a braking action without skid, hence the required bandwidth is within the range of a few Hz.

---

4 The AFCS typically performs several functions depending on the mode in which it operates (AP, CAS, etc.).

5 A pure delay of 0.25 s introduces a 90° phase shift at 1 Hz, a phase shift equally produced by a system of the second order with a natural frequency of 1 Hz.

By pushing the limits of human control, the development of computer-controlled commands can significantly increase the performance of an aircraft, and at different levels:

– for the *flight envelope* the reduction of margins is permitted by the speed of monitoring and limiting reaction, timing and decoupling. This allows, for example, Limiting structural loads due to maneuvering and gusts (Manoeuver Load Alleviation or MLA and Gust Load Alleviation or GLA) or to increase passenger comfort by reducing the impact of accelerations;

– for the *stability* of the aircraft (Stability Augmentation System, or SAS) through the introduction of correctors in the flight control laws that prevent, for example, aerodynamic coupling, like the Dutch roll;

– for the *flight qualities* in order to improve aerodynamic efficiency, for example, by lowering the ailerons (aileron droop) when the flaps are deployed in the landing phase;

– for the *dynamics* regarding the generation of flight control setpoints, for example, hyper-maneuverability, provided that the actuators have sufficient bandwidth.

I.1.5. *Facilitating development and integration*

The use of computers to develop or process orders issued from the human pilot provides flexibility in the development of an aircraft. It is, for instance, possible to change very quickly the control laws of the actuators in response to the orders of the pilot in order to assess them, or even to emulate the performance of another plane.

Note also that the transmission of information in electrical form releases many geometric integration constraints in terms of the routing for control signals within the airframe.

The list of needs is of course huge as it depends on the application. Using military aircraft as the example, we could cite many improvements on survivability[6] and the simplification of operational support.

---

6 Aircraft combat survivability (ACS) is defined as the capability of an aircraft to avoid or withstand a man-made hostile environment.

I.2. Functions and architecting

Actuation can be perceived in different ways depending on the perspective we adopt. Concerning the architecture, concepts and sizing, we draw largely on a graphic representation that allows us to highlight the processes and the quantities by which they operate. We may, for example, adopt the generic form shown in Figure 1.5 of Volume 1, or choose other more appropriate graphics to represent either the vision for signal or the vision for power. An intermediate representation that describes both aspects of signal and power is often adopted as a compromise.

I.2.1. *Signal vision*

If we take the point of view of the control, it more often than not shows the actuation in the form of block diagram, like the one in Figure I.1 below.

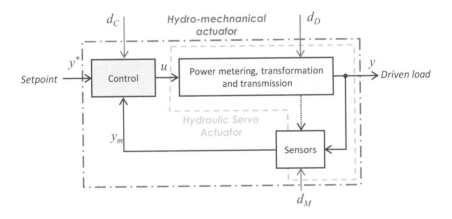

Figure I.1. *Example of a block-diagram representation of an actuator according to a signal vision*

The values circulating on the pathways connecting the blocks are considered as pure signals. The controlled variable is generally a position (e.g. flight controls) or a force or pressure (e.g. braking). The inputs are either functional stimuli (e.g. the order y^* issued by a pilot) or disturbances d. These disturbances can be external (e.g. the measurement disturbance d_M, the variation in power supply d_D (or power disturbance) and the quantization noise of control signals d_C). They can also result from coupling produced by the power transmission between the "blocks" (e.g. opposing force of a control surface on its actuator or flow consumed by the set of brake pistons). Thanks to the measures y_m, the control develops the orders u to ensure that the output variable y is consistent with the orders y^* (tracking function) and is insensitive to various disturbances d (rejection function). In general, the power source and the power transfer do not appear explicitly on this representation.

It is important to note that the boundary of an actuator is not universally defined. In terms of aircraft, the actuator is most often associated with a physical unit that the aircraft manufacturer integrates between the airframe and the driven load. It will be seen later that the development of an actuation modifies the boundary of the actuator as defined by its physical interfaces. This is illustrated in Figure I.1 by the dotted lines. For example, for hydro-mechanical actuators that receive a mechanical position setpoint, the "control" block is purely hydro-mechanical. This is opposite to conventional hydraulic servo-actuator for which the Electro-Hydraulic Servo Valve (EHSV) current, that is to say the control signal u, is calculated in computers[7] that are often still located in the cockpit. In the actuators which are electrically signaled only (*full SbW*), the control and measurement signals are sent in electrical form.

I.2.2. *Power vision*

If we instead seek an energy vision, the graphical representation emphasizes the energy transfers between the various components (blocks in the block diagram) of the actuator, since the energy flows from sources (inputs) to the users (outputs). This is interesting as it explicitly distinguishes the visions for signal and power. For example, arrows in bold lines can be used to illustrate power flows. In Figure I.2, this distinction is reinforced using a half arrow in reference to the Bond graph [KAR 00]. We further augment readability by using different colors depending on the physical field.

---

[7] This is no longer the case on the Airbus A350 for which the command u is produced in electric form in the actuator which directly receives the set point y^* (see photograph in Figure 1.6 of Volume 1).

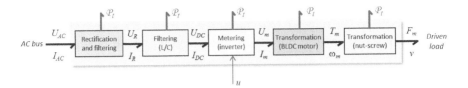

Figure I.2. *Example of block-diagram representation of an electro-mechanical actuator according to a power vision. For a color version of this figure, see www.iste.co.uk/mare/aerospace2.zip*

The power bonds carry both power variables (see Table 1.3 of Volume 1), for example voltage U and current I, pressure P and volume flow rate Q or force F (or T) and velocity v (or ω). We may also consider these with respect to their time integral (e.g. the position in the mechanical domain). From a functional point of view, it is often necessary to define causalities on these links, that is to say, what is the variable imposed on the block involved (the cause) and what is the variable that responds to the link (the result). This distinction may be apparent, for example, the position control where the actuator's function imposes the load's position, and in turn produces an opposing force. However, the choice of causality to adopt the representation is not always that simple. It depends in particular on the level of detail: does an electric motor impose torque in response to the applied current or speed in response to the applied voltage? It further depends, for example, whether or not we consider the inertia of the rotor or the current feedback loop.

As the focus is on the exchange of power, it is useful in some cases to represent the heat flux $\mathcal{P}_t$ to the atmosphere. This provides a balanced power representation when performing energy audits. Similarly, it would be necessary to represent the mechanical power links associated with force flow from the structure and from the load [DAU 14], which would provide a mechanically balanced representation (these connections are not shown in Figure I.2 in the interests of clarity). The *Power-by-Wire* (PbW) actuators are powered by a source of electrical power. For these actuators, the need to use balanced representations in force and in power is much stronger than for the hydraulically powered actuators. Indeed, in these PbW actuators, we will see that the rejection of the heat generated by power losses, reaction forces and inertial effects all impact heavily on the architecture and overall sizing. In practice, it is unfortunately difficult to produce diagrams which completely reflect these balances (in the sense of effort, energy, etc.) while maintaining readability.

It is equally interesting to explicitly highlight the signals that act on/control the power transfer (e.g. the input current of a servovalve or the command in the form of a pulse width modulation, which is applied to an inverter).

In practice, the distinction between signal and power, however, is not unique. It depends on the system under consideration, the engineering task and the level of detail of the study. For example, the electrical control of a flight control actuator can be viewed as a signal, without power transfer. Indeed, the power required to control a servovalve (some 10 mW) is negligible when compared to the power transmitted by the actuator on a moveable surface (a few kW to tens of kW). Similarly, at the level of the flight control system, the dynamics of the servovalve (a few tens of Hz) can be neglected when compared to the bandwidth of the position control (a few Hz). By contrast, from the perspective of the computer that issues the orders to the actuator, it may be necessary to take into account the input electric power and the dynamics of the servovalve coils.

I.2.3. *Basic functions in actuation*

Understanding the needs and the evolution of actuation functions is facilitated by firstly considering human piloting in view of flying rules. It is quite possible to represent the general architecture in the form of a block diagram, as shown in Figure I.1. It is considered that the output is the actual attitude and trajectory of the aircraft. The functional input is the y^* desired by the pilot. The disturbance input d alters the actual output (e.g. wind gusts affecting the flight controls, imperfections of the runway on the landing gear, etc.). The pilot therefore aims to achieve a tracking function (the actual output of the aircraft follows the desired input by the pilot) and a rejection function (the actual output of the aircraft is immune to disturbances acting on the aircraft).

In this representation, it is therefore possible to distinguish:

– a comparator function and generation of the command u. The pilot compares the desired attitude and trajectory y^* of the aircraft with his perception y_m of the actual output of the aircraft. He creates the command u to apply the inceptors (control stick, pedals, etc.) in order to reduce the error between the desired change and the actual output;

– a measurement function on the actual response of the aircraft. The pilot perceives the aircraft's attitude and trajectory through the sensors which act as extensions of his senses (sight, the feeling of acceleration, noise, etc.);

– an actuation function in the strictest sense of the meaning. In manual piloting, that is to say without any amplification of force, the muscles of the pilot carry out the function of the actuator. They develop the power required to move parts of the aircraft via the inceptors. This power is then transmitted mechanically (rods, levers, pulleys, cables, etc.) or hydrostatically for non-assisted braking (master cylinder, tubes and cylinders receptors).

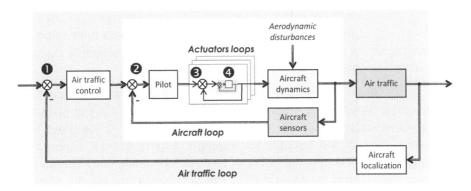

Figure I.3. *The different feedback loops for an aircraft*

This breakdown generates several observations:

a) when talking about a closed-loop control system, there is implicitly a control function and a power amplification function. The power is provided by a source outside the considered system, although it does not explicitly appear on the graphic in Figure I.1.

b) in practice, there are usually several nested feedback loops, as shown in Figure I.3: the aircraft in air traffic control ❶, the trajectory and attitude of the aircraft ❷, the actuators of the aircraft ❸ and finally, the internal feedback loops to the actuator ❹ (e.g. internally from the feedback of a servovalve spring or the feedback from the motor power drive of a brushless electric motor). The more internal feedback loops, the higher the dynamics (e.g. 1,000 Hz for a current loop, for a servovalve 70 Hz, 3 Hz for an actuator, etc.).

c) the signals which are propagated between the blocks of the block diagram in Figure I.1 are multidimensional. For instance, regarding the control interfaces, we can indeed mention the stick, rudder pedals and the elevator trim wheel. To activate movable surfaces, we can distinguish the actuators for roll, pitch, yaw, the horizontal stabilizer, etc. For sensors, there are, for example, the gyrometers, the airspeed sensors, the position sensors on movable surfaces, etc.

d) some of these "blocks" generate strong couplings between the various input and output signals. These couplings may be undesired, e.g. a yaw induced by aileron deflection on an airplane; or by increasing the collective pitch of the main rotor of a helicopter. These couplings can also be functional, e.g. at the level of a mixer for controlling the main rotor actuators of a helicopter; or a command law for the uncoupling integrated into the flight control computer.

e) for the flight controls, it will later be seen that the CAS and SAS produce other links in the internal feedback loops. The CAS augments the pilot's commands as issued to the actuators so as to improve the maneuverability of the aircraft, for example a turn without slipping or skidding. The SAS rejects the effect of rapid aerodynamic disturbances and improves the stability of the aircraft. The AP substitutes for the pilot to generate the flight control commands to the actuators, as per the setpoint of the aircraft loop, according to the desired flight path as provided by the Flight Director (FD).

In the end, it is clear that the success of a flight mission depends on several generic actuation functions. To facilitate reading and reduce complexity, the rest of the book will distinguish these functions as either an aspect of "signal" or an aspect of "power". However, it will also repeatedly show how these two aspects can be coupled, specifically for all the intermediate power levels between the computers and the load, e.g. concerning the power command of a Direct Drive Valve (DDV); or the supply of a solenoid valve. These aspects are illustrated by the following two practical examples.

I.2.3.1. *Example 1: Flight control actuator of the Northrop B2 Bomber*

Figure I.4 shows a representation, as taken from a control point of view, of one of the two actuators associated with the active–active mode for operating a flight control surface of a Northrop B2 Bomber [SCH 93].

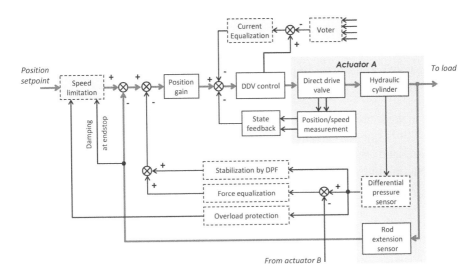

Figure I.4. *Links and control functions to be performed electrically on a flight control actuator of a Northrop B2 Bomber*

In this example, it is possible to identify several functions related to the active mode of the actuator[8] that must perform an electrohydraulic position control of the load. This major function is highlighted by the thick arrows. It uses the measurement of the rod extension that is provided by a sensor integrated into the actuator. The Direct Drive Valve (DDV) is itself position controlled by a state feedback structure of order 2 which uses the measurement of the power valve opening and valve opening rate. The dotted boxes in the graphic represent important additional functions utilizing the measurements of current and differential pressure for the following purposes:

– limiting the rate of change in the setpoint position to the approach of endstops and when the forces transmitted are too great;

– equalizing the currents produced by the quadruplex commands through a voter;

---

[8] It is noted that the term actuator is used in this book to denote the physical unit which is installed on the aircraft to operate the load. In this case, the actuator does not include any electronics.

– stabilizing underdamped hydro-mechanical mode by Dynamic Pressure Feedback (DPF);

– equalizing the force applied to the moveable surface by both actuators (A and B) attached to it.

When we consider that the control electronics is not integrated within the actuator, it is not difficult to imagine the number of electrical connections to establish, and the number of operations required to perform all these functions, including those to excite the sensors and demodulate their output signals.

I.2.3.2. Example 2: rudder actuator for the Boeing B777

The rudder actuator of a Boeing B777 as shown in Figure 1.12 illustrates the importance of power management. It is a hydraulically supplied and electrically controlled actuator that uses an electrohydraulic servovalve to feed power accordingly. On the hydraulic diagram of Figure I.5, it appears that the actuator can operate in three modes depending on the supply of three solenoid valves[9]:

– damping mode: when any of the solenoid valves are energized. The actuator behaves functionally as a damper because the mode selection valve isolates the two hydraulic ports of the cylinder from the servovalve control ports and interconnects them by hydraulic restriction;

– active mode: when the bypass solenoid valve is energized to enable the normal mode, regardless of the orders issued to the other two solenoid valves. In this case, the actuator is position controlled. The mode selection valve connects the two use ports of the servovalve to both hydraulic ports of the cylinder;

– free mode: when the damping solenoid valve is energized and the bypass solenoid valve is not. The actuator is functionally disengaged from the load: the mode selection valve isolates the two hydraulic ports of the cylinder from the servovalve control ports and interconnects them without introducing hydraulic resistance.

---

9 In the absence of specific standards to represent the three-state mode selection valve with respect to its control inputs, a personal representation was adopted.

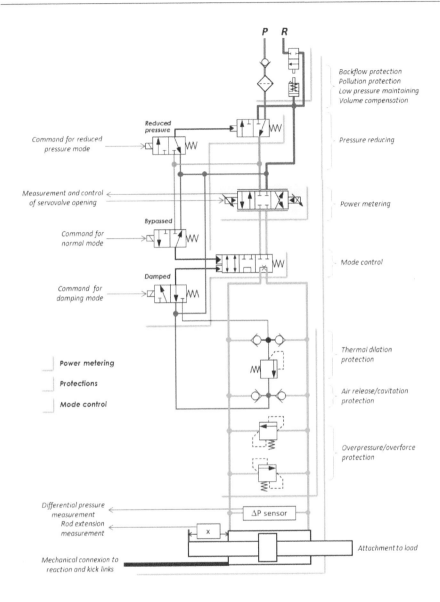

Figure I.5. *Electrical, mechanical and hydraulic interfaces of a rudder actuator for a Boeing B777 (according to [SHA 15]). For a color version of this figure, see www.iste.co.uk/mare/aerospace2.zip*

When we examine the left side of Figure I.5 in view of the required number of electrical interfaces that appear, we consider again the amount of electrical connections to establish, the operations required to achieve the closed-loop control function, and the mode selection function. Note also that the electrical signals for control and mode selection shall also transmit power: a few tens of mW for the driving current of the servovalve and some W for the solenoid valves. This is a characteristic of hydraulically supplied actuators. It can be seen as an advantage as the actuator requires no electrical supply to control the power management functions. Conversely, it may be perceived as a drawback when regarding the transmission of data.

I.2.4. *Hydro-mechanical actuators*

Figure I.6 will be used later to assist in understanding the evolution of the actuator to more or fully electric. In a simplified format, it outlines the power and control architectures of hydro-mechanical actuators, that is to say those supplied with hydraulic power (constant pressure P and variable flow Q) and receiving the reference signal y^* in mechanical form, take as an example the actuator shown in Figure 7.12 of Volume 1.

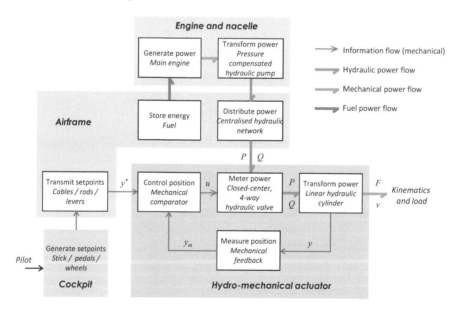

Figure I.6. *Joint representation (signal/power) of hydro-mechanical actuation. For a color version of this figure, see www.iste.co.uk/mare/aerospace2.zip*

The whole information chain is implemented through mechanical technology. The actuator locally performs the function of position control. Only the setpoints are conveyed from the cockpit to the actuators which are located closer to the loads to be moved.

The power chain is hydro-mechanical. The power is generated centrally at the engines of the aircraft and put at the disposal of the actuators at constant pressure through a distribution network.

I.2.5. *More or total electric actuation*

Numerous functions related to actuation were first carried out without recourse to electricity. For automatic flight, we can, for example, cite the Sperry GyroPilot autopilot whose principle for the roll axis is given in Figure 1.8 of Volume 1. The move to more or fully electric affects all functions. However, we find that the transition is (or has occurred) more or less quickly as per the needs of the various types of aircraft, the technological maturity and the criticality of the actuating function.

The PbW actuators are interfaced at the power level with the onboard electrical power networks. It then becomes possible to suppress the hydraulic power generation/distribution networks and their specific disadvantages, which are mentioned in paragraph 1.3.4 of Volume 1. For their part, the SbW actuators are interfaced electrically at the signal level, which eliminates the mechanical information chains, as those are slower, less accurate and more difficult to integrate into the airframe.

It thus appears that the advancement towards "X-by-Wire" takes place mainly in the form of incremental innovation: the changes are introduced very gradually by way of conventional solutions as they are validated and accepted. It is a strong characteristic of aerospace and we must master the risk associated with the evolutionary process, particularly in light of life cycle duration (typically 30 years of production and 30 years of operation) and because of the requirements of availability and reliability. Maintainability is also a major constraint: the skills, best practices and tools must be mastered and made available anywhere aircraft are operated[10]. As discussed in the following chapters, disruptive innovation is very rare. Moreover, different time scales naturally emerge when comparing the evolution of military aircraft, commercial aircraft, helicopters and launchers.

---

10 For helicopters or business aircraft, the majority of operators only operate a few aircraft, sometimes thousands of kilometers away from the major maintenance centers.

1
Electrically Signaled Actuators (Signal-by-Wire)

The objective of this chapter is to present the architectures for the transmission/processing of information associated with actuation functions. The "flight control" application is used as the basis for this exploration because it is much more complex, and in many ways richer than other applications (e.g. landing gear, engines, etc.). This chapter will explain the evolution of flight controls from purely mechanical to all electrical signaling. To illustrate the importance of the electric transmission of information and power in modern aircraft, it is best to start off by stating a few orders of magnitude:

– the Airbus A380 has nearly 500 km of electric cables. The 120 miles of electrical cables for the Boeing B787 has a mass of 4 tons;

– Daniel [DAN 07] considers that on current long range aircraft, we can count 13,400 functional power lines (for both power and signal) for a total length of 240 km and with a total mass of 1,800 kg (we have to add 20% mass for connectors and fasteners). The part pertaining to the transmission of information is typically comprised of 7,800 lines, 170 km long and with a mass of 700 kg. This consists of 5,400 mono wire cables for analog and discrete signaling, with 2,050 twisted pair cables and 350 coaxial or quadraxial cables;

– for helicopters, [VAN 07] indicates that the Sikorsky S-92 is comprised of 295 kg of electrical cable, including 193 kg for information transmission and with 1,200 cable/aircraft interfaces. The mass of electrical cables increases to 612 kg on the CH-53K model, representing 3.5% of its empty weight.

1.1. Evolution towards SbW through the example of the flight controls

As indicated at the end of the previous chapter, the electrification of the information chain took place "gradually". Whatever the application (fixed wing or rotor wing; commercial, military or private), when the electric information chain was first introduced, the mechanical information chain was maintained as a standby. The latter gradually disappeared, as redundancies applied to the power information chain allowed for the new alternative to reach the level of reliability required. The richness of this evolution, as illustrated by flight control, is analyzed below. It also attempts to embrace other actuating functions associated with the landing gear (steering and braking) or the engine (control of air and fuel flows, geometry and thrust reversers).

1.1.1. *Military applications*

The trends towards the uptake of more electrical information chains were initially driven by the need to autopilot flight in situations with zero visibility (military aircraft) or without an actual pilot (missiles and space launchers). For military aircraft, reducing vulnerability and increasing maneuverability have also been major reasons for the electrification of both the information and the control chains. This is illustrated in Table 1.1, which has mainly been sourced from [RAY 93].

Year	*Manufacturer model*	*Advancements (from the previous model)*
1943	Boeing B17E	3-axis autopilot with electrical output, mechanical summation on pilot commands by an electro-mechanical actuator
1959[*] 1964[**]	Dassault Mirage IV	Electrohydraulic elevons, analog electric signaling with mechanical backup
1964[*] 1967[**]	General dynamics F-111	Electrohydraulic spoilers
1974[*] 1978[**]	General dynamics F-16	Analog quadruplex FbW to increase relaxed static stability Quadruplex electronics with middle value logic Side-stick
1974[*] 1982[**]	Panavia Tornado	Triplex digital electronic FbW and mechanical backup channel

1978* 1983**	McDonnel Douglas F/A-18	Quadruplex digital electronic FbW Separation of electrical and hydromechanical fault detection and isolation Backup mechanical signaling for pitch control
1974* 1986**	Rockwell B1-B	Combination of full authority SCAS and purely hydromechanical control chains Structural Mode Control System (or SMCS)
1979* 1984**	Dassault Mirage 2000	Electrohydraulic actuators, and quadruplex analog FbW, electrical backup
1986* 2001**	Dassault Rafale	Electrohydraulic actuators, triplex digital FbW and analog backup Side-stick
1989* 1997**	Northrop B2	Quadruplex digital electronic FbW 4 Actuator Remote Terminal (or ART) in the wings, communication by multiplex bus Performance level kept in case of failure of an actuator electronic channel by changing the loop control gains of remaining healthy channels

*First flight, **Entry into Service (EIS)

Table 1.1. *Evolution towards an all-electric information chain for the flight controls of military aircraft*

1.1.2. *Commercial aircraft*

Table 1.2 shows the incremental evolution of flight controls for commercial aircraft, using European aircraft as an example. Note that it took a cumulative total of over 40 years to completely replace the mechanical information chain with all-electric information chain.

Year	Manufacturer model	Advancement (from the previous model)
1969* 1975**	Sud Aviation - British Aircraft Corporation/ Concorde	Analog FbW with mechanical backup Analog electrical signaling between cockpit and actuators
1972* 1974**	Airbus/A300B	Analog FbW for 12 non-essential functions Analog Electrical signaling between cockpit and actuators Position servo control performed by computer in the cockpit

1982[*] 1983[**]	Airbus/A310	24 FbW actuators controlled by five computers Removal of low speed ailerons (roll control assisted by spoilers at low speed) Introduction of electrically signaled trim for ailerons and rudder
1987[*] 1988[**]	Airbus/A320	FbW on 3-axis, 7 digital computers Mechanical signaling between the yaw damper and the rudder actuators Backup mechanical signaling (pseudo FbW) for rudder and Trim Horizontal Stabilizer (THS) Introduction of the side-stick
1991[*] 1993[**]	Airbus/A330 340	5 digital computers Removal of any mechanical signaling for the rudder control (A340-600) with resort to an electrical analog Backup Control Module (or BCM)
2005[*] 2007[**]	Airbus/A380	6 digital computers Removal of the last mechanical backup signaling for THS control (*Full* FbW)
2013[*] 2015[**]	Airbus/A350	6 digital computers Position servo-control at actuators level (Actuator Control Electronics or ACE)

[*]First flight, [**]Entry into Service (EIS)

Table 1.2. *Evolution towards an all-electric information chain for the flight controls of European commercial aircraft*

The advances in information chains that are mentioned in this table resulted in a significant and continuous improvement on performance [VAN 02]. Of particular note:

– between the A300-B4 and the A310, there was a mass reduction for flight controls of approximately 300 kg, predominantly due to the removal of low speed ailerons permitted by the complementary use of spoilers for low speed roll control;

– between the A310 and the A320, a mass of 200 kg through improved flight controls, an increase in security by the protection of the flight envelope, a reduction of the pilot load that negated the need for a flight engineer (two-crew operation) and a reduction of dynamic airloads through the Load Alleviation Function (or LAF);

– between the A300 and the A340, a 45% mass reduction of the rudder actuators (of identical specifications) and a further mass reduction of about 50 kg on the A340-500/600 from the removal of the mechanical controls and the associated actuators for the yaw damper, in favor of an electric control (see Figure 1.20);

– on the A380, a mass saving through reduced stability margins for longitudinal balance (area of the trim horizontal stabilizer reduced by 10% thanks to the introduction of FbW for pitch control, including for the backup channel);

– on the A350 [AIR 13], an increase in the aerodynamic efficiency of the wing due to the Differential Flap Setting (or DFS), the Variable Camber (or VC) and the Adaptive Dropped Hinge Flaps (or ADHF).

1.1.3. *Helicopters and compound helicopters*

On airplanes, lift is provided to the wing due to relative airspeed. Furthermore, a given attitude control function (at least for roll or pitch) is often provided by several moving surfaces, themselves driven by a redundant actuation system (e.g. two separate actuators as on commercial airplanes or a dual body actuator as with many military aircraft). Given the dynamics involved, it is generally accepted to temporarily lose the positioning function of a control surface in the event of actuator failure. We can even accept to permanently lose the positioning function, provided that actuators respond as required to a failure (depending on the need: fail-passive, fail-neutral or fail-freeze; see Volume 1, Chapter 2).

On helicopters, lift and attitude are controlled by the cyclic pitch and collective pitch of the blades. The angle of attack of each blade must be controlled absolutely at each moment: as such any transient loss of the positioning function is not permitted[1]. Moreover, in the event of a failure of the flight control functions, the boundaries of the flight envelope are reached significantly faster than on a plane.

---

1 For example, we typically tolerate a runaway at maximum speed during of 20 ms or less, which produces a displacement of rotor actuator rod not exceeding 5 mm.

All these considerations explain why the transition of flight control signaling, from mechanical to electrical, takes so long to occur. This is illustrated in Table 1.3 which indicates that the development of helicopter flight controls has taken nearly 70 years [STI 04]. Faced with the criticality of helicopter flight controls, FbW was gradually phased in but commands remained transmitted mechanically. At any time, this design enabled a safety pilot to take the control through a backup mechanically signaled channel in case of failure of the FbW channel.

Year	Manufacturer model	Advancements (from the previous model)
	Increase in mechanical stability and dynamics	
1941	Young patent [YOU 45]	Hover stabilization by Young-Bell bar
1953	Sikorsky/H03-S1	Improved longitudinal stability through action on the blades pitch as a function of aerodynamic forces generated on air foils
1954	Bell/47	Additional vertical acceleration feedback by inert mass
1964	Bell/H13	Mechanical first-order lead filter to improve the roll rate response
	First electronic autopilot	
1950	Piasecki/HUP-1	Introduction of the electronic autopilot
1952	Sikorsky/S-56	Improved response to gusts and controllability
1960	Sikorsky/S-58	"Hands-free" hovering
	Research in the United States to improve reliability and maturity	
1973	Boeing/CH47B	Demonstration of digital FbW with redundancy management (TAGS project)
1975	Boeing/XCH-62	Response to commands and stability augmentation level made dependent on the phase of flight (HLH project)
1986	Sikorsky/UH-60A	Optic/electric flight control with evaluation of side-sticks and improved agility (ADOCS project)
1992	Sikorsky/ S-76 Shadow	Transition during the takeoff and landing phases
	Research in Europe and Canada	
1979	Bell/205	FbW demonstration flight, NRC, Canada
1985	BO/105 ATTHeS	FbW flight simulation, DLR, Germany

Electrically Signaled Actuators (Signal-by-Wire) 7

1991	Aérospatiale/ AS 365 Dauphin	FbW demonstration flight, Aérospatiale, France
2001	Bell/412 ASRA	FbW demonstration flight, NRC, Canada
2002	Eurocopter/EC135	Fly-by-light demonstration flight, DLR, Germany

Table 1.3. *History of the development of electric flight controls for helicopters and compound helicopters*

Table 1.4 confirms the difficulty and time spent for moving FbW designs into service: in 2015, the only models with FbW in mass production were the NH Industries NH90 and the Boeing V22 Osprey; while several programs have since been abandoned.

Year	Manufacturer model	Characteristics
1975* 1984**	Boeing/ AH-64A Apache	Attack Helicopter Mechanical signaling with backup FbW on each axis (*Back-Up Control System* or BUCS)
1989* 2005**	Boeing/ V22 Osprey	Multi-role combat tiltrotor
1992* 1994**	Mc Donnel/ MD 900 Explorer	Light helicopter Analog FbW for vertical stabilizer (Vertical Stabilizer Control System or VSCS)
1995*1 2003*2 2007**	NH Industries/ NH-90	Middle class military helicopter First mass-produced military helicopter equipped with FbW (quadruplex) series
1996* Abandoned 2004	Boeing/Sikorsky RAH-66 Comanche	Combat Helicopter, FbW (triplex) without mechanical backup
1998* Abandoned 2014	Sikorsky/CH-148	Multi-role naval helicopter
2003* 2018** ?	Agusta Westland/ AW609	Civil transport tiltrotor equipped with FbW (triplex)
2015* 2017** ?	Bell/ 525 Relentless	Multi-role civil transport helicopter. First commercial helicopter equipped with FbW (triplex) without mechanical backup.

* First flight, ** Entry Into Service (EIS), [1] mechanical mode only, [2] full FbW mode

Table 1.4. *Examples of FbW industrial helicopter and compound helicopter programs*

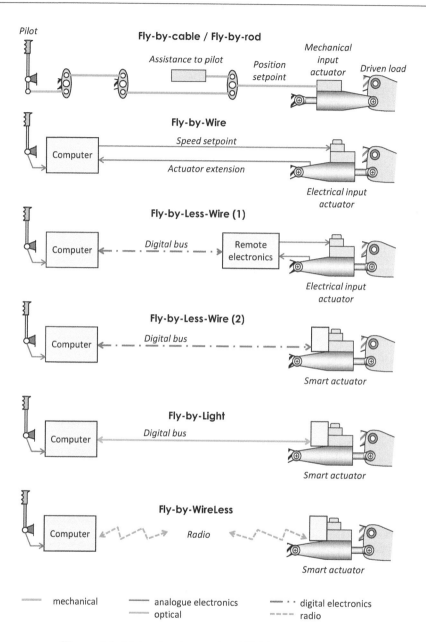

Figure 1.1. *Major developments of flight control signaling*

1.2. Incremental evolution from all mechanical to all electrical

Under the signal aspect, the evolution of flight control can be shown in Figure 1.1. Each architecture, presented here in a simplified way, will be discussed in further detail in the following sections. In practice these changes occurred in small increments, which are not all shown by this figure:

– there were no actuators on the first flight controls;

– between the Fly-by-Cable and Fly-by-Wire, we could have distinguished the introduction of actuators with dual inputs, both electrical and mechanical. First, the electrical input appeared alongside the servovalve, to directly inject into the actuator the augmentation or autopilot commands. Next, the electrical input became the main input, with the mechanical transmission path being used as a backup before being gradually phased out.

1.2.1. *Exclusively mechanical signaling*

In the designs of conventional flight controls, the pilot's commands move the load exclusively transmitted in mechanical form: here we might speak of "Fly-by-Cable" or "Fly-by-Rod" versus "Fly-by-Wire". Figures 1.2–1.4 provide examples of such architecture as employed in military aircraft (McDonnell Douglas F-15, EIS 1976), commercial aircraft (Airbus A310, EIS 1983) and helicopters (Sikorsky S-76, EIS 1979).

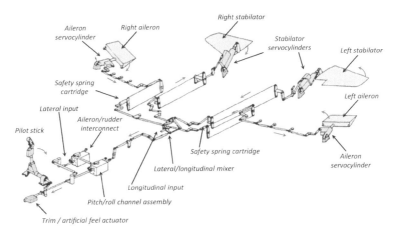

Figure 1.2. *Roll control for the McDonnell Douglas F-15 fighter (http://www.f15sim.com). For a color version of this figure, see www.iste.co.uk/mare/aerospace2.zip*

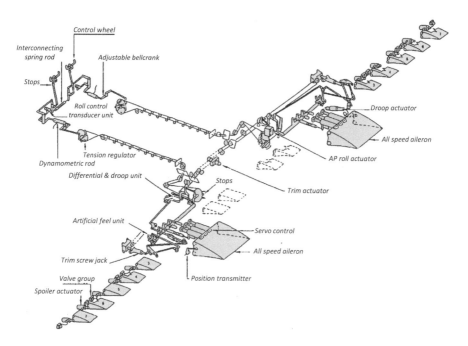

Figure 1.3. *Roll control for the Airbus A310 [VAN 02]. For a color version of this figure, see www.iste.co.uk/mare/aerospace2.zip*

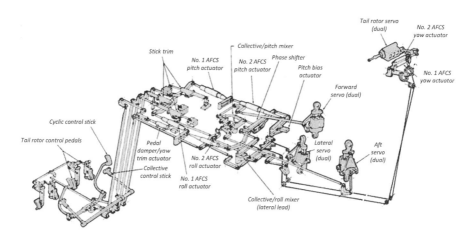

Figure 1.4. *Flight controls of the Sikorsky S-76 helicopter (© I. Sikorsky historical archives). For a color version of this figure, see www.iste.co.uk/mare/aerospace2.zip*

While different aircraft types have different strengths, their architectures share many similarities, particularly those listed functions and their gradual introduction. The addition of these functions is realized by mechanical summation through two main principles:

– The position (or movement) summing. For the example given in Figures 1.5(a) and (c), this is performed by crossbar, or hydraulic or electro-mechanical actuators integrated in series. That is to say, each of these channels adds a displacement on the mechanical signal chain. A feared event would be a break in one of these channels, as in this case the remaining mechanical parts can move independently of one another;

– The force summation. As shown in Figure 1.5(b), it is realized by an arm whose position is given by one input and propagated to others. A feared event would be the seizure of a pathway, thereby freezing the entire mechanical chain. To avoid this situation, we usually install spring rods, Figure 1.5(d), on the chains in series, so as to decouple the channels, thereby overcoming force. Note the example in Figure 1.3, at the interconnection of the two pilots' control sticks, or again for the transmission of mechanical position setpoints to the three aileron hydraulic cylinders from a single source movement.

1.2.1.1. *Primary piloting functions*

The pilot uses conventional inceptors, control stick/control wheel and rudder pedals to change the position of the loads to be actuated (in these examples, flaps, swashplate of the main rotor and/or the tail rotor).

The pilot's orders are transmitted mechanically to the loads to be actuated through connecting rods, cables, levers, cranks or even flexible cables (ball bearing cables) as employed in the McDonnell Douglas F-15 Eagle fighter. When needed, further elements are installed to ensure tension regulation if the transmission is achieved by cable. Figure 1.6 shows the flight control cables from the Aero Spacelines Super Guppy cargo aircraft. For loading and unloading, the hinged nose of the plane that constitutes the cargo door is pivoted about a vertical axis. Thereby it requires all the flight control cables to be disconnected.

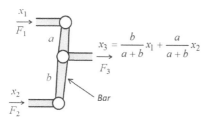
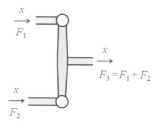
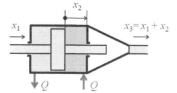
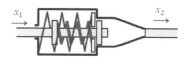

Figure 1.5. *Summation principles of mechanical transmission. For a color version of this figure, see www.iste.co.uk/mare/aerospace2.zip*

Figure 1.6. *Flight control cables of the Super Guppy cargo plane*

If needed, the mechanical commands are combined in a mechanical and passive manner, using the position-summing concept. Thus, coupling or decoupling functions are produced. Figure 1.3 shows an example of droop actuators that steer the ailerons down to improve aerodynamic efficiency when the flaps are extended. Figure 1.4 provides a further example through the mixer that develops swashplate commands, based on the collective and cyclic orders generated by the pilot.

1.2.1.2. *Augmentation and automation of control commands*

Functions for the augmentation and automation of control commands are added through the auxiliary actuators operating on the mechanical information chain. As shown in Figures 1.2–1.4:

– AutoPilot (AP) or Automatic Flight Control System (AFCS) actuators;

– compensation actuators (trim) that balance the aircraft for a particular flight phase when pilot commands are at neutral;

– actuators for specific functions such as Stability Augmentation System (or SAS). For example, the yaw damper on the yaw control command to avoid the Dutch roll;

– actuators for limiting the structural loads such as the rudder travel limiter as a function of true airspeed;

– actuators for specific functions such as the Control Augmentation System (CAS) in manual control.

Figure 1.7 shows how the auxiliary actuators are integrated into the mechanical signal chain for a conventional flight control axis[2]. It is represented in the generic intermediate form, addressing both aspects of signal and power.

---

[2] As shown in Figure 1.8, of Volume 1, the first functions (e.g. primary flight controls) of this type were initially implemented without any electrical device.

14   Aerospace Actuators 2

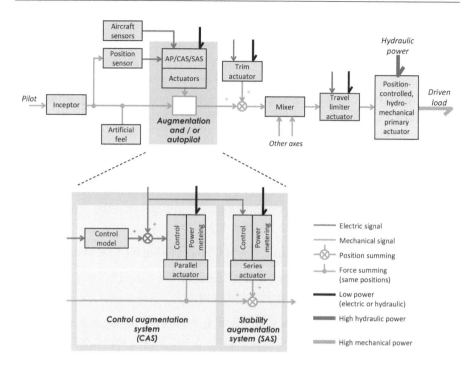

Figure 1.7. *Generic architecture, signal and power, for a conventional flight control axis. For a color version of this figure, see www.iste.co.uk/mare/aerospace2.zip*

The bottom half of the diagram shows an example for the implementation of an augmented or autopilot system. In practice, this implementation is slightly different depending on the type of aircraft (fixed or rotary wing aircraft):

1) Stability Augmentation System function. The SAS actuator is mounted geometrically in series on the mechanical signal chain to produce a summation of position. It must be quick enough to respond to aerodynamic disturbances. Its total stroke is only a few percent of that of the mechanical chain, which allows for manual control in the event of SAS malfunctions. It is located downstream of the artificial feel system. Therefore, it produces orders that propagate towards the main actuator, which do not alter the position of the pilot inceptors: its action is thus invisible to the pilot who does not perceive it.

2) Control Augmentation System function. The CAS actuator is geometrically mounted in parallel: it acts by imposing the position command

relative to the airframe to the primary actuator. Its action is thus directly reflected at the level of pilot inceptors and consequently perceived by the pilot. The actuator has a higher authority, several tens of percent, of the functional stroke of the mechanical chain. In the event of the function failing, the pilot must be able to counter the forces produced by the actuator which should therefore include a release system. Conversely, in the case of a malfunction of the mechanical drive chain upstream of the actuator, the actuator can be used as a backup to pilot the aircraft via the commands issued by the autopilot. The force summation that can be seen as position sharing is implemented in most cases, as shown in Figures 1.2–1.4, by means of spring rods.

3) Compensation function. The trim actuator performs an average compensation that only requires a very low bandwidth. It is realized through position summing on the mechanical chain so as to ensure that the aircraft is aerodynamically balanced. This reduces the average forces that the pilot needs to generate and maintains the inceptors at neutral position, regardless of the stabilized flight phase. The trim actuator must be disconnected in the event of malfunction.

4) Load limiting function. The travel limit actuator bounds the permitted stroke according to flight conditions. This ensures the integrity of the structure by ensuring that the aerodynamic loads do not exceed the limit values used for sizing.

The power that must be generated by auxiliary actuators is generally low (a few tens of N, a few cm/s). On the other hand, their criticality is often moderated because they can be disengaged or surpassed by the pilot. Hence why it is possible to achieve this in electro-mechanical form, as is generally the case for commercial aircraft. When they need to be more dynamic, these actuators are usually electrohydraulic (see Volume 1): the hydraulic power metering function is performed by an Electro-Hydraulic Servo Valve (EHSV) or a Direct Drive Valve (DDV). In addition, the hydraulic technology can likewise realize a compact and easy way to accommodate these disengagement functions. This type of actuator is found on helicopters for the AFCS or SCAS functions. Furthermore, the CAS and SAS functions can be integrated as a single function (SCAS, AP or AFCS) using a single actuator mounted in series on the information chain. This function can be made redundant. For helicopters, integration into the airframe is facilitated by grouping these auxiliary actuators on a single physical unit in charge of the three (or four) axes of motion: roll, pitch, yaw (and collective pitch), as shown in Figure 1.8 for the Eurocopter

Tiger SCAS. Note the presence of the dual input servovalve, mechanical (position control) and electrical (SCAS function).

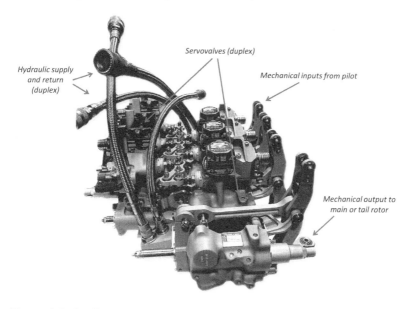

Figure 1.8. *Auxiliary actuator SCAS group of the Eurocopter Tiger helicopter*

Figure 1.9 shows the CAS servovalve control diagram for the pitch axis control of the McDonnel Douglas F-15 fighter. Note the large number of inputs and analog functions to be performed electronically to implement the CAS function, whose authority is ±10° of deflection for the stabilator. As shown in Figure 1.2, the CAS function is applied by an additional electrical input on the mechanically signaled flight control actuator.

1.2.1.3. *Reduction of piloting forces*

When the forces to be applied on the driven load exceed human muscle capacity, a force amplification function has to be installed. Although some regional aircraft still use an aerodynamic assist in the form of servo tabs or control tabs, this amplification function is most often accomplished with actuators, referred to as primaries, which are nonetheless still overwhelmingly hydromechanical. Thus, primary flight control actuators (primary servo controls) are inserted in series between the pilot inceptors and the load to be moved, as close to the latter as the integration within the airframe allows. They are supplied by hydraulic power and they servo control the position of the driven load in response to the pilot demand, i.e. transmitted by the mechanical information chain. Therefore, these are hydromechanical actuators, which is to say, hydraulically powered and mechanically signaled, as shown in Figure 1.10. The power architecture and geometric integration of such a hydromechanical position servo actuator are discussed in Chapter 7 of Volume 1. For each critical function, the actuation is generally rendered redundant (triplex actuation, active–active–active, as shown in Figure 1.3, dual–duplex actuators, active–active, as shown in Figure 1.4) to achieve the required level of reliability.

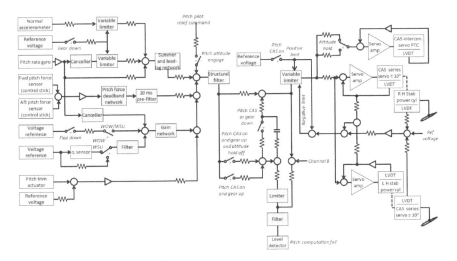

Figure 1.9. *Analog control diagram of the CAS servovalves for the pitch control of the McDonnel Douglas fighter F-15, (http://www.f15sim.com). For a color version of this figure, see www.iste.co.uk/mare/aerospace2.zip*

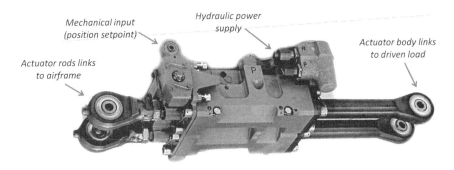

Figure 1.10. *Dual-parallel hydromechanical aileron actuator for the Falcon 900*

As the flight control actuators are irreversible, they do not reflect any force feedback to the pilot (to nearby forces due to unwanted effects in the mechanical signaling chain). It is therefore necessary to give the pilot a force feel[3] based on the position command which he sent to the actuators. For this, passive devices (artificial feel unit in Figure 1.3, pedal damper in Figure 1.4) create opposing forces on the mechanical information chain.

1.2.2. *Fly-by-Wire*

The transmission of information in mechanical form offers very good reliability, thereby often negating the need for the redundant path. Unfortunately, it has many flaws, whose importance often increases with the size of the aircraft and the number of loads to be moved. For example, the Boeing 747-400 has more than 20 flight control actuators some of which are located some 60 m away from the cockpit. It is thus clear that the effects of dilation, structural elasticity, friction or play in the transmission of command signals strongly penalize the accuracy, dynamics and geometric integration of the mechanical information chain. Moreover, the incremental evolution of adding auxiliary actuators on this command chain quickly reaches its limits in matters of authority, coordination of loads to be moved, mass, protection of the flight envelope and others. Airbus considers in particular that the introduction of FbW on its A320 has produced a real gain in mass of 200 kg. A similar gain was obtained by introducing the Load Alleviation Function (LAF).

As electrical signaling becomes more reliable, the more it becomes tempting to remove auxiliary actuators (AFCS, AP, SAS, CAS). It even becomes possible to completely remove the mechanical information chain if all of its functions can be achieved by signaling the primary actuators electrically. This leads to the Fly-by-Wire solution where the primary hydromechanical actuator is replaced by a Hydraulic Servo Actuator (or HSA)[4].

---

3 By his sense of touch or muscle reaction, the human being finds it easier to control forces rather than positions. In practice, this is often the mechanical impedance (relation force/position) of the actuated load that allows the human being to control positions. When this is not enough, a force feel system shall be introduced, for example through the installation of a simple spring.

4 An example for the power architecture of a hydraulic servo actuator for flight control can be seen in Figure 7.6 of Volume 1. Sometimes also called servo-hydraulic actuator or electro-hydraulic servo actuator.

1.2.2.1. *Full Fly-by-Wire*

Figure 1.11 is equivalent to Figure 1.7, when all the information is transmitted to/from the actuators in electrical form. We can note the complete disappearance of the mechanical information chain, the recourse sensors to measure the pilot's commands and the state of the actuator as well as the control and monitoring of the actuator mode (e.g. active/ disconnected/dampened/blocked). The aircraft feedback loop is shown to indicate, at least for monitoring, the position of the load to be actuated. Note that in practice, this position is somewhat different from the position provided by the actuator sensor. This comes from the kinematic gain variation of the actuator/load transmission, of the joints backlash, and of the deformation of solids under load.

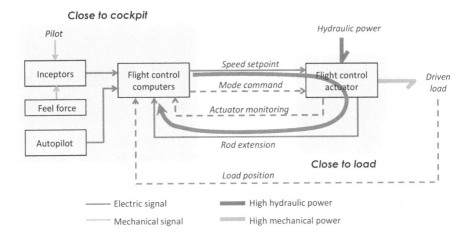

Figure 1.11. *Generic architecture, signal and power, of an electrically signaled flight control axis. For a color version of this figure, see www.iste.co.uk/mare/aerospace2.zip*

Figure 1.12 shows one of three full FbW actuators for the positioning of the rudder for the Boeing B777. It highlights interfaces for power (mechanical and hydraulic) and signals (electric). Internal power interfaces between electrical and mechanical domains (servovalve, mode selection solenoid valves or pressure reduction) are also displayed. The reaction and kick links (Figure 7.14 Volume 1) are not mounted.

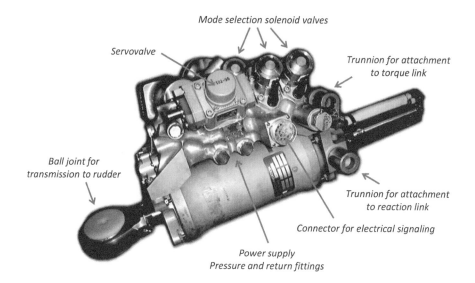

Figure 1.12. *Full FbW actuator for the rudder control of the Boeing B777*

The signal architecture shown in Figure 1.11 calls for several major remarks:

1) In conventional architectures, the primary actuator receives a position setpoint. It is in charge of locally performing, hydromechanically, the servo control position of the load according to this setpoint. This solution has the advantage of not being particularly vulnerable. However, it reduces the opportunities of advanced correction in the control loop [MAR 99]: overlapped/underlapped/dual slope valve or even the more complex Dynamic Pressure Feedback (DPF). On the one hand, in the FbW architecture of Figure 1.11, the actuator receives the servovalve current (a few mA) which is functionally representative of the desired load speed. Because the coils of the servovalve have a low impedance, another advantage of this solution is that it is not sensitive to noise or electromagnetic interference. Furthermore, as the power control of the servovalve (a few 10 mW) does not require high power electronics, its reliability is increased. In addition, the fact that the computers need to be located nearer the cockpit, in pressurized and temperature controlled avionic zones, helps to further increase reliability. However, when we consider that there may be several tens of meters between the computer and the actuators, it is understood that the transmission of commands or

sensor feedback is potentially vulnerable: depending on the type of fault, which can produce jamming, free or erratic movement, directly affecting the load to be actuated.

2) In a full FbW architecture, the interest in the removal of a number of mechanical components is offset by the number of electrical signals to be conveyed, typically between 15 and 25 wires per actuator channel (e.g. four wires to control the servovalve, three times two wires to control the mode solenoid valves, six wires for the ram position sensor, six wires for the position sensor of the mode selection valve). On a commercial aircraft, it is typically considered that there are between 350 and 450 electrical wires connecting the computers located in the cockpit to the flight control actuators [KUL 07]. FbW on a helicopter, such as the NH-90, for which electrical signaling is quadri-redundant, takes on average 60 wires for each of the four flight control actuators, totaling more than 200 electrical wires. Later on, it will be seen how the number and length of wiring has been reduced in recent programs.

1.2.2.2. Pseudo Fly-by-Wire

When the level of reliability achieved by full electric signaling is not enough[5], the solution is to double the electrical channel with a mechanical channel, hence introducing a dissimilarity that is beneficial for reliability. In the vast majority of cases, the electrical channel is the normal operation working chain and the mechanical channel is used as the backup. A counter example is the AH-64D Apache helicopter, in which the electric flight command is used as a backup in the event that functionality of the mechanical channel is lost. In this case, the pilot must land the helicopter as soon as possible because the reliability of the electric channel is not sufficient to continue the mission safely.

The pseudo FbW was implemented on the flight commands of the supersonic Concorde; for the yaw control and for the control of the trim horizontal stabilizer on the Airbus A320 and A330-200; to control the stabilizers of McDonnell F-18 fighter [HAR 83]; or as an intermediate step to full FbW for the NH90 helicopter, as shown in Figure 1.13. In this last example, the hydraulically powered primary actuator has both an electrical input and a mechanical input.

---

5 For example, a failure rate lower than 10^{-9}/flight hour is typically required for a flight control axis (see Volume 1).

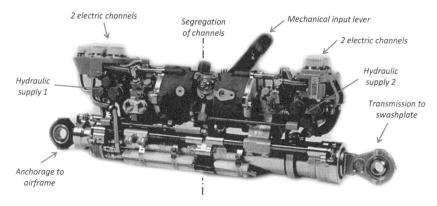

Figure 1.13. *Main rotor actuator of the NH90 helicopter, the pseudo FbW version used in the flight tests for development*

1.3. Challenges associated with electrical signaling

1.3.1. *Electrical interfaces*

Compared to conventional flight controls, the electrification of actuators imposes reliable interfaces to or from the electrical domain: power interfaces to meter or manage hydraulic power in the actuators from an electrical signal; human machine interfaces to collect the orders of the pilot; sensors to measure the physical quantities of the actuator and the aircraft.

1.3.1.1. *Power interfaces*

For hydraulically powered actuators, the interface between electric signaling and hydraulic power domains is performed by the servovalve that has been studied in detail in Volume 1, Chapter 5. From a very weak power of electric command (a few tens of mW), the servovalve meters high hydraulic power (tens of kW), giving it a gain in power in excess of a million. The power is metered with good dynamics (some 10 Hz bandwidth), good linearity (combined linearity/threshold/hysteresis of only a few % of the nominal range) and with an excellent capacity to reject external disturbances, in particular vibration and temperature.

The initial trend for military and space applications was to develop highly specific servovalves for each program. As such they could respectively accommodate internal mechanisms to stabilize the position control or satisfy the reliability requirements (power management and reconfiguration as a function of the desired response to a fault). The introduction of augmentation or autopilot functions created the dual input servovalves, as shown in Figure 1.14, of the Eurocopter EC225. The auxiliary electrical input issued by the functions of AP, AFCS or SCAS is superimposed on the primary mechanical input command issued by the pilot.

Figure 1.14. *Dual input servovalve for the SCAS of the Eurocopter EC225 helicopter*

1.3.1.2. *Interfaces to measure physical quantities within the actuator*

To overcome the great spatial extent of the electric position control loop in Figure 1.11 and to avoid recourse to an electric position sensor, the first FbW controls were implemented locally, that is to say at the level of the actuator, hence the position feedback was purely hydromechanical. Figure 1.15 shows this principle through the example of the Thrust Vector Control (TVC) of the main engines for the NASA Space Shuttle.

The electric signal input of the actuator thus consigned the position setpoint, which was locally transformed into electromagnetic force. The comparison function of the position servo loop was then realized by comparing this electromagnetic force to the elastic force produced by a spring cage as a function of the relative position between the actuator's rod and body. As of the 1960s, this type of solution is used to control the thrust vector of the space launcher programs for Gemini, Saturn and Space Shuttle, as well as on the SAS actuators of the General Dynamics F111 bomber.

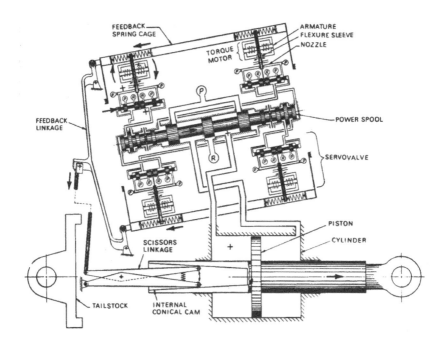

Figure 1.15. *Purely hydromechanical feedback loop actuator for the FbW TVC actuator of the main engine in the NASA Space Shuttle (© Moog Inc.)*

Later on, it came to implement the FbW position feedback electrically. Resistive potentiometric position sensors were discarded due to the presence of sliding contact. Friction impacted the lifespan (although the average position is quasi-invariant, the average speed is quite high), reduced reliability and generated measurement noise. Among the concepts for non-contact position sensors, the uptake of the variable differential transformer [MEA 13, NOV 99] has quickly become a standard feature for several reasons:

– it is well suited to the measurement of the absolute linear position, including for long strokes for this application; it is called Linear Variable Differential Transformer (LVDT). If necessary, it can also be extended for the measurement of absolute angular position, becoming a Rotary Variable Differential Transformer (RVDT). This is frequently used in the steering controls of auxiliary landing gear. As the measurement range of RVDTs is only a few tens of degrees, they can be combined if necessary with a backlash free mechanical gearbox;

– the sensor body that contains the coils can be easily sealed and made resistant to the operating pressure on the actuators (e.g. 350 bar);

– the core need not be guided precisely within the body: the diametral clearance may be several tenths of a mm;

– the ratiometric demodulation (see below) allows the sensitivity of the electrical output to temperature and voltage excitation to be greatly reduced.

An LVDT is a variable transformer. The body includes a primary excitation coil and two secondary measurement coils. The mechanical position to be measured is imposed by a non-magnetic rod to a ferromagnetic core. The displacement of the core relative to the body modifies the magnetic coupling between the primary coil and the two secondary coils, as shown in Figure 1.16.

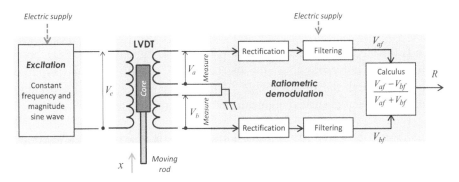

Figure 1.16. *Functioning principle of LVDTs and ratiometric demodulation*

The primary coil is fed by a sinusoidal voltage (a few kHz) of constant amplitude (a few V): the carrier. As a result, the current that arises generates a variable magnetic flux which induces voltages V_a and V_b in the two secondary coils. These voltages have the same frequency as the carrier but are amplitude modulated by the relative position of the core in regard to the body. To demodulate these voltages with a view to produce the electric signal that is representative of the position to be measured, the carrier must be eliminated by rectifying and low-pass filtering of voltages V_a and V_b. Two concepts are mainly implemented:

– the *synchronous demodulation* uses only the voltage difference $V_a - V_b$ which is obtained by connecting the two secondary coils in series and in phase opposition. Thus limiting the number of wires of the sensor to four (two for excitation and two for measurement). To be precise and to demonstrate good linearity in the vicinity of the center position, this solution needs to compensate for phase shifts (introduced by the unwanted capacitances) between the voltages of the coils, which requires the use of active electronic circuitry.

– *ratiometric demodulation* is preferred in aviation because it eliminates the phase effects, the thermal drift and the influence of the excitation magnitude. To this end, the LVDT transducer is designed to keep the sum of the voltages V_a and V_b constant across the entire functional stroke. After rectification and filtering, these tensions can be directly processed by computer to form the ratio:

$$R = \frac{V_{af} - V_{bf}}{V_{af} + V_{bf}} \quad [1.1]$$

which is the electrical image of the core position relative to the LVDT body, a null ratio corresponding to the centered position. This type of demodulation has the disadvantage of using separate V_a and V_b voltages, which requires the electrical connection of six wires (three coils).

For their robustness to the environment and their accuracy, LVDTs are also used to measure other physical quantities where they can be turned into position: a force sensor by measuring the deformation under load; a pressure sensor by measuring the hydrostatic force on the surface of a membrane.

1.3.1.3. *Human–machine interfaces*

The removal of the mechanical chain transmitting the commands of the pilot offers new opportunities for the ergonomics of the cockpit. However, aircraft manufacturers have taken this benefit in different ways:

1) Central column vs. side-stick

The first option is to retain the control stick (or the central column with its roll control wheel) in line with conventional flight controls. In this way, the human–machine interface is unchanged for a pilot who progresses to a FbW aircraft. The second option is to introduce side-sticks in order to clear the space in front of the pilot. This improves vision from within a glass cockpit and/or allows for the possibility of touchscreens.

2) Passive vs. active central column or side-stick

The first option privileges reliability. It consists of using a passive column or side-stick, that is to say, only capable of opposing a predetermined force to the action applied by the pilot. Its main drawback is the inability to reproduce the mechanical link between the pilot and co-pilot commands. This feature enables one pilot to perceive the actions of the other and is nevertheless very useful both for training and also to detect instances of double pilotage. Equally, the removal of this feature makes it impossible for the pilot to feel, through the control stick, the commands generated by the autopilot or the vibrations generated by a shaker announcing the risk of stalling. Finally, the mechanical impedance of the control stick, that is to say, the force felt by the pilot in response to his actions, cannot be dynamically modified to suit the flight conditions and the specific ergonomic choices of the aircraft manufacturer.

The use of an active column or mini stick, which involves actuators capable of producing a driving force on the stick, remedies these disadvantages. However, in so doing it raises strong challenges for performance and reliability, especially in instances where side-sticks must be combined. An intermediate solution is to use a semi-active design. It only generates an opposing force, according to the desired mechanical impedance, that can be controlled electrically.

The supersonic Concorde was equipped with analog pseudo FbW on three axes. The control stick with the rolling wheel and the conventional rudder had been preserved. However, as the servo controls were irreversible, two force-controlled electrohydraulic actuators, shown in Figure 1.17, were associated in a parallel according to an active–passive architecture on each axis, in order to make the pilot feel an opposite force depending on flight conditions.

Before the introduction of FbW on its A320 in the mid-1980s, the European manufacturer, Airbus, had already conducted tests in 1978 on Concorde, and then again on a modified A300-B2 [ZIE 85]. The findings led them to retain the solution of a passive stick with position control (centimetric stroke) rather than force control. When the autopilot is engaged, the side-stick is blocked in the central position. In case of need, the pilot may override this blockage by applying sufficient force to disengage the autopilot and take manual control. In manual operation, normal flight control laws[6]

---

[6] Other control laws are activated in case of faults: for example, alternative law (the protections are removed but the computers still apply correction in the actuators' position control loops) and direct law (the pilot's commands are transmitted unmodified to the flight control actuators).

consider action on the side-stick as a request for acceleration. The flight control computer has the authority to limit the request so as to keep the aircraft within its flight envelope: limiting load factor (e.g. between +2.5 and −1 G); attitude protection (e.g. a maximum of 30° for pitch-up and 15° for pitch-down); angle of attack; and overspeed protection.

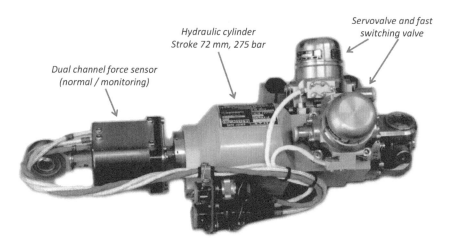

Figure 1.17. *Artificial feel actuator for the Concorde*

Since their introduction on the Airbus A320, the side-sticks, shown in Figure 1.18, have made significant progress in terms of integration [BET 15]. It is estimated that the choice of a side-stick over a central column saves 60 kg on the A320. The efforts and progress in integration thus save an estimated additional 5.5 kg on the A380, a further 3 kg on the A400M and another 2 kg on the A350 XWB.

Bell has chosen to expand the use of the side-stick for the 525 Relentless helicopter, which will be the first civil full FbW helicopter; with a side-stick on the left-hand for the cyclic controls and another side-stick on the right-hand for the collective pitch control.

For its part, Boeing chose to keep both conventional central columns, mechanically coupled, with the introduction of FbW on its B777. In this aircraft, the flight envelope remains under the complete authority of the human pilot: unlike the choice of the European aircraft manufacturer, commands are not limited by the flight control computers which only produce "dissuasive" limits through alerts and tightening of the control stick as the limits of the flight envelope are approached, so that the pilot forces the airplane to depart from its normal flight envelope.

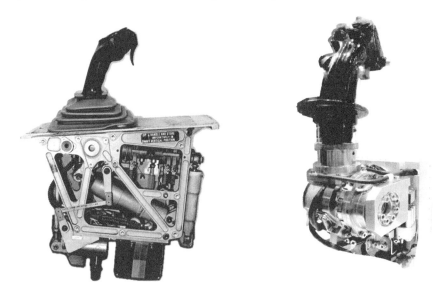

Figure 1.18. *Side-stick Left: passive (A320, courtesy of SwissTeknik LLC); Right: active (KAI T-50)*

Active side-sticks are already implemented for military applications, such as the Lockheed Martin F-35 multi-role fighter and the Sikorsky CH-53K heavy helicopter. According to [WAR 15], the feedback seems to have convinced aircraft manufacturers of the maturity of this option, which was selected for the Embraer KC390 medium-transport military aircraft and the Gulsftream G500/G600 business aircraft. As the certification requirements are much more stringent for civil than for military aircraft, side-sticks, should be, for example, dual–duplex (*dual* type for command/monitoring architecture on each channel and *duplex* for the presence of two channels associated in parallel), with two different processors and quadri-redundant sensors. Depending on the manufacturer's choice, both the control sticks of the pilot and the co-pilot can be connected "electronically" to replicate the existing conventional mechanical conjugation between the sticks and to reflect the actions of the autopilot.

1.3.2. *Evolution of the control and information transmission architectures*

As so often is the case with innovation, the reliability of FbW poses a dilemma for designers. First, greater use has to be made of the redundancies in order to meet the reliability requirements. Second, the multiplication of redundant channels needs to strike a balance with the mass and the increase in the complexity of information transmission, which in turn has an effect on overall reliability. It is a very gradual evolution that occurs over several years or even decades.

1.3.2.1. *Centralized analog FbW for actuator control*

At the beginning of the introduction for FbW (e.g. for the Concorde or the Dynamic General-F16), the actuator feedback loop was performed by analog electronic circuits situated in a less hostile environment, most often in the cockpit, in a pressurized and temperate zone.

1.3.2.2. *Centralized digital FbW for actuator control*

Analog circuits were then replaced by digital computers (e.g. Airbus A320), and still located with the avionics in the cockpit. In this architecture, both the issuing of the flight control commands and the closed-loop position control are performed far from the actuators. This is highly disadvantageous regarding the number and lengths of electric wires, as outlined above.

1.3.2.3. *Remote and mutualized digital FbW for actuator control*

The first step to achieve a more distributed architecture was taken in the mid-1990s, on the Northrop B-2 bomber [SCH 93] or the commercial aircraft Boeing B777 [YEH 96]. The actuator position control and monitoring functions are decentralized and integrated in electronic units; for example, eight Actuator Remote Terminals (or ART) for the B-2 or four Actuator Control Electronics (or ACE) for the B777. The ACE or ART interface with the digital avionics on the one hand (ARINC databus MIL-STD-1553 or ARINC 629, respectively) and with the analog electric signaling of the actuators on the other. There are still no electronics at the actuator level. As the position command is nearer to the actuators, the Flight Control Computers (or FCC) mainly transmit the actuator position commands in digital form. This is an advantageous solution for the required data rate. It is in fact much lower than would be necessary if the computers were having to "close" the position control loop and transmit the command to the actuator's servovalve.

Figure 1.19 shows a simplified diagram of the digital FbW architecture of the remote and integrated actuators' control for the Boeing B777.

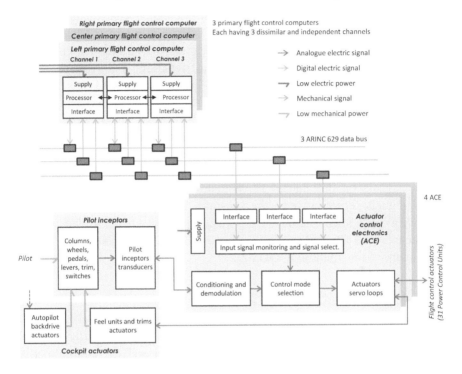

Figure 1.19. *Signal architecture with ACE in the Boeing B777. For a color version of this figure, see www.iste.co.uk/mare/aerospace2.zip*

1.3.2.4. *FbW with smart actuators*

The lessons learnt enabled more reliable installations of the electronics in harsher environments (temperature, pressure, humidity, vibration, electromagnetic interferences, lightning, high intensity magnetic fields). Therefore, it became possible to integrate the electronics within the flight control actuators to make so-called *smart actuators*.

One of the first applications for commercial aircraft concerns the trim horizontal stabilizer of the Airbus A330/A340, which integrated three digital electronic controls communicating, via the ARINC 429 databus (see section 2.2.3 of Chapter 2), with the flight control computers. On the A380, the electronic controls of the Electro-Hydrostatic Actuators (EHA) are used as backup (see Chapter5) and are designed to receive the same type of command signals as used by conventional hydraulic servo-actuators for the normal mode: a rated current of a few mA which is representative of the desired speed for the load to be moved. The integration of these electronics into the actuator for the local control of the motor (commutation, current and speed loops) was another step towards smart actuators. Yet another step was taken on the recent Boeing B787 and Airbus A350 XWB, as all flight control actuators, conventional or PbW, incorporated an electronic module (REU for Remote Electronic Unit or RAE for Remote Actuator Electronics). This module is responsible for carrying out the functions for position control and for the monitoring of the actuator. Equally, it also appears as an interface between the digital data on the side of the avionics databus and the analog electric signals on the side of the actuator (e.g. commands for control of power, signals returning from the sensors). The connection to the databus requires a very small number of wires (typically 4), compared with the number of wires (typically 15–25 depending on redundancies) that connect the actuator to this module [GOD 02].

1.3.3. *Reliability and backup channels*

To reach the levels of reliability required for FbW, it is usually necessary to install a backup channel. As shown in Tables 1.1 and 1.2, an intermediate solution, widely used in the electrification of the information chain, is to keep a mechanical channel as backup (e.g. the mechanical input on the actuator in Figure 1.13). The total removal of the mechanical linkages between the cockpit and the actuators imposes new constraints on matters of independence and dissimilarity for the entire backup chain (pilot inceptors, closed loop control, power sources for command).

1.3.3.1. *Backup with fluidic technology*

In the early 1970s, as part of the American research program FLASH (Fly-by-Light Advanced Systems Hardware), one of the concurrent goals was concerned with solutions towards fluidic computation [CYC 81]: the entire backup chain to transmit and process information between the pilot and the actuator was designed to be performed by pneumatic fluidics. This line of investigation was however quickly abandoned as the immunity of electrical solutions to electromagnetic interferences became significantly improved.

1.3.3.2. *Analog electrical backup*

One widely used solution consists of implementing the backup channel in analog electric technology. A good example is the evolution of the yaw control between the Airbus A340-200 and the A340–600 (see Figure 1.20).

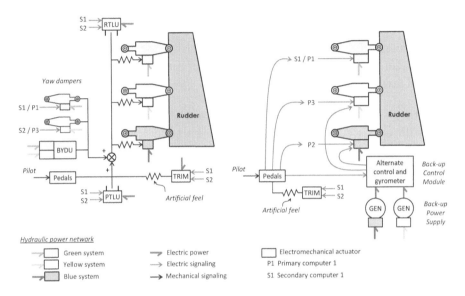

Figure 1.20. *Evolution of yaw control from hydromechanical actuation (Airbus A340-200, left) to electrohydraulic actuation (Airbus A340–600, right). For a color version of this figure, see www.iste.co.uk/mare/aerospace2.zip*

On the -200 version, the actuation architecture was largely similar to that of the Airbus A320. This is presented in the left half of Figure 1.20:

– each of the three control actuators is mechanically signaled, hydraulically powered and locally realizes a closed-loop position control with a moving body design. The three actuators are associated with force-summing configuration and operate in the active–active–active mode to position the rudder;

– the mechanical commands defining the position setpoint of the three flight control actuators are identical. They are issued from the pilot or autopilot. The commands for the yaw damper and yaw trim that are produced at the tail of the aircraft are also added to this. Position commands are transmitted to the actuators through spring rods to ensure independence between channels in case one actuator fails;

– the mechanical yaw damper commands are generated by two electrically signaled hydraulic cylinders associated in force-summing configuration. As a backup, a Yaw Damper Backup Unit (BYDU) is added to the A320 design. It performs the yaw damping function in the event of a double failure of the yaw damper actuators (given that the A340 is more disposed to Dutch roll);

– to bound the aerodynamic structural loads, the position demand is limited according to the airspeed (Pedal Travel Limitation Unit or PTLU and Rudder Travel Limiting Unit or RTLU).

On the A340-600, the flexibility of the airframe required an increase in bandwidth and accuracy for yaw control. This led to the removal of the mechanical signaling channel in favor of full FbW. This architecture is shown at the right-hand side of Figure 1.20:

– the three fixed-body actuators are electrohydraulic, that is to say, hydraulically supplied and power metered by servovalve;

– besides the position closed-loop control function, the primary and secondary flight control computers are also in charge of the damping functions for yaw and travel limitation. They receive the position commands from the pilot in electric form through the pedal inceptors, and generate the control current for each actuator's servovalve;

– the trim command remains electro-mechanical but is sent back to the cockpit;

– a backup channel is added to the yaw control. The constraints of autonomy, segregation and dissimilarity are cleared by means of an analog Backup Control Module (BCM) with its own gyroscopes, its own pedal sensors, as well as its own Backup Power Supply (BPS). This redundant supply produces electrical power through two generators, which are driven by hydraulic motors from centralized sources of pressure;

– a force equalization strategy between the three actuators, not indicated in the figure, is implemented in the position control of the actuators in order to limit force fighting.

1.3.3.3. *COM/MON architecture*

A major challenge in meeting reliability requirements is the ability to detect and isolate faults that may occur in the computation of the command within an actuation system, so that they do not result in flight control failure.

The COM/MON (Command/Monitoring) architecture briefly covered in Chapter 2 of Volume 1 and detailed in Figure 1.21, is a widely used solution [TRA 06]. The two channels COM and MON are simultaneously active. In terms of software they are completely independent, segregated and dissimilar. In principle, the command channel (COM) generates the commands for power metering and management that are transmitted to the actuator. The monitoring channel (MON) validates or inhibits their transmission to the actuator, and enables (or not), the active mode of the actuator (see Figure 7.6 of Volume 1).

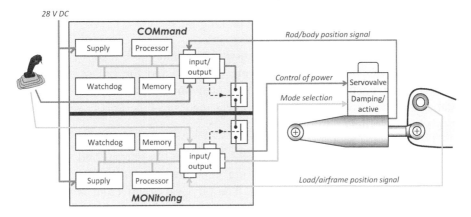

Figure 1.21. *COM/MON architecture for controlling the actuators (according to [TRA 06])*

1.4. The example of landing gears

The actuating functions related to landing gears (steering, braking, etc.) have broadly followed the same incremental changes of the flight controls. This development, similar to that observed in the automotive industry, can be illustrated through the example of braking. The left-side of Figure 1.22 shows a typical conventional braking system of an aircraft. For simplicity, a single circuit and a single wheel are represented and the parking brake is not mentioned. Each of the blocks is associated with different evolution steps:

– initially, the braking was purely physical, the forces being transmitted by a mechanical cable between the pilot and the wheels. Then the cables were replaced with hydrostatic power transmission, thereby overcoming the friction and compliance inherent to the previous transmission cables;

– the forces exerted by the pilot were assisted through the introduction of an external hydraulic power source;

– the anti-skid function was added to improve steerability and braking performance. It works by reducing the intensity of braking if excessive wheel/ground slippage is detected. Note that the anti-skid function, as shown in the figure as electronically controlled, appeared very early in purely hydromechanical form;

– finally, automatic braking (autobrake) helped to reduce the pilot workload for the management of the braking intensity.

It is clear that the assistance, anti-skid and autobrake features appear as add-ons on the original chain linking the pilot and the brakes. However, unlike flight controls, they are not located on the information chain, but rather the power chain, inserted as power metering functions in series (anti-skid) or in parallel (autobrake).

The right-hand section of Figure 1.22 shows the partial and simplified braking system of an airliner, taking full advantage of SbW (N.B. the redundant paths are not illustrated). The physical architecture is simplified and its integration facilitated. The control, brake assist and anti-skid functions report to a Braking and Steering Control Unit (or BSCU) and a single power metering device (pressure servovalve). The hydraulic lines between the cockpit and the braking system are replaced by electrical wires. A recent example of this illustrates well the scalability of SbW: combined with modular avionics architecture, the A380 or A350 Airbus can offer the *Brake to Vacate* function (or BTV). This feature aims to increase airport traffic by clearing the runway faster after landing. It is realized by automatically metering the braking of the wheels according to the position of the aircraft in relation to the ground in order to vacate the runway at a predetermined exit.

Electrically Signaled Actuators (Signal-by-Wire)   37

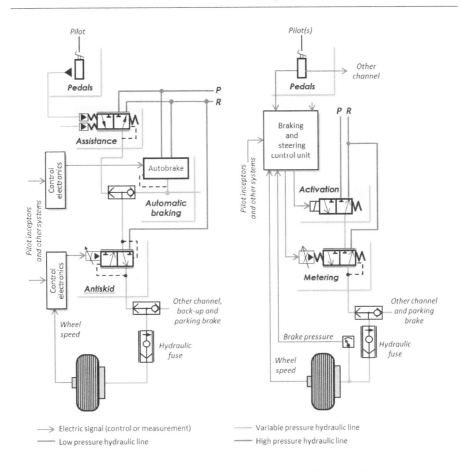

Figure 1.22. *Evolution of braking. Left: "conventional" brake with antiskid and autobrake (from [CRA 01]); right: fully integrated electrohydraulic brake. For a color version of this figure, see www.iste.co.uk/mare/aerospace2.zip*

2

Signal-by-Wire Architectures and Communication

As it progresses, Signal-by-Wire (SbW) has increased the total amount of data being exchanged, as well as the number of communication units and the number of interactions taking place between these units. This signal traffic has had to grow with improved flexibility, scalability and reliability of complex and critical systems. Two major changes have helped achieve these goals: the first being the massive use of digital data networks and second being the introduction of integrated modular avionics. Given that this greatly reduces the number of electrical wires needed for the transmission of information, it is sometimes referred to as Fly-by-Less-Wire (or FbLW) to point out this difference with previous generations.

With full SbW architecture, data is exchanged in electrical form between sensors, actuators, computers and human–machine interfaces. At the end of this chapter, it will be shown that other methods for the transmission of data exist aside from electrical ones (e.g. optical fiber or radio signals). Regardless of the medium being used, data transmission typically employs the elements shown in Figure 2.1. First, the data to be transmitted are encoded and formatted (e.g. in a serial digital form). The data are then introduced onto the medium (modulation) by the transmitter according to the selected standard. Thus the source essentially creates the information medium (power, light, carrier frequency, etc.). At the destination, the data is retrieved from the medium (demodulation) via the receiver and reconstructed for use. At the signal level, formatting, modulation and demodulation functions can be perceived as the interfaces between the components exchanging data, and the transmission lines carrying them.

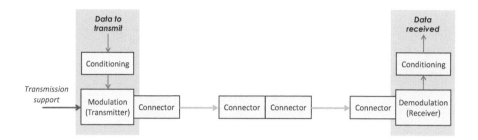

Figure 2.1. *Data transmission*

2.1. Architectures

As mentioned in the previous chapter, the incremental evolution of avionics first entailed the addition of extra components to existing architectures, which naturally led to more distributed architectures. Subsequently, a more comprehensive view of the transmission and processing of information between systems led to federated architectures, followed by more integrated modular architectures [HAM 03, KOP 04, MOI 13].

2.1.1. *Federated architectures*

The first avionic architectures appear as an aggregate of systems, each with their own computing resource: analog followed by digital. In this distributed architecture, each resource is designed to realize a given function, whether it be hardwired or programmed. Hence it exists as a dedicated physical unit, often as a Line Replaceable Unit (LRU), which incorporates its own control unit, power supply and input–output interfaces. Each system exchanges as many signal inputs and outputs with other systems as it needs. In a federated architecture, as shown in Figure 2.2, the exchanges between systems are shared with one or more data buses.

The properties of a federated architecture directly result from each system being developed and operated almost entirely on an independent level:

1) Advantages:

The scope of responsibilities for each supplier is clearly established. Development requires little interaction between the system suppliers. As there is little interdependence between the systems, the risk of error propagation via the transmission of data is therefore limited. With regards to aircraft manufacturers, reliability and availability are essentially system-specific and validation is simplified. It is often possible to reuse, with minor modifications, systems which have already been developed for previous applications.

2) Disadvantages:

Computation resources are often underutilized and certain functions must be duplicated unnecessarily on multiple systems. With regards to aircrafts, all this unnecessary duplication strongly penalizes the attributes of bulk, mass, power consumption and cost (acquisition and maintenance). It is moreover difficult to implement more integrated functions. Scalability is low because whenever new features are intended to be added at the system or aircraft level, a reexamination process is often required to sensibly review the concept and its proposed integration in detail.

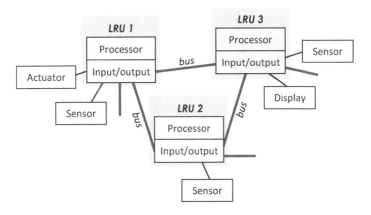

Figure 2.2. *Federated architecture*

2.1.2. *Integrated modular architectures*

It becomes very difficult to improve system performance if each system is developed independently, as it is already highly optimized. In order to produce significant performance gains at the level of the entire aircraft, it would be better to take into account all the interactions and couplings between systems and determine how to make better use of these resources (processors, memory, etc.). This objective is unfortunately difficult to achieve with a federated vision of avionics architecture, as it is based on the parallel implementation of several systems, which are themselves highly independent.

Conversely, Integrated Modular Avionics (IMA) is seen as a unique system, and therefore its hardware and software resources are shared by integrated, common platforms (see Figure 2.3).

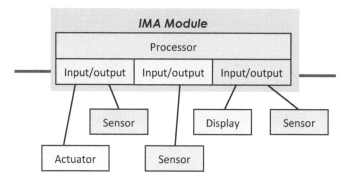

Figure 2.3. *Integrated modular architecture*

For integration, a processor is no longer dedicated to a single application but instead performs several system applications and the data transmissions between units are achieved through a high-speed multiplex data network. For modularity, computers are standardized and are not specifically developed or dedicated to a particular application. By configuration, they can allocate a part of their resources to particular system applications. The installed computational capacity is better exploited, and the operation is optimized, and the cable lengths for data transmission are minimized. For example, the integrated modular avionics of the Boeing B787 runs 70 applications at various levels of criticality and has simultaneously eliminated 100 LRU, versus a federated architecture [DOW 09]. With IMA, the disadvantages of a federated architecture disappear but in so doing loses the benefits associated with independence: allocation of responsibilities, management of error propagation, and integration and validation are therefore subtler. These drawbacks are greatly reduced by adopting the modular structure in Figure 2.4. The ARINC 653 and DO 297 standards define the development processes and the solutions implemented for such an architecture:

– a module houses a processor, a partitioning core and partitions dedicated to each application;

– responsibilities are clearly distributed: to the supplier of the application and the platform source (processor/core) and to the integrator responsible for the entire system;

– the system integrator allocates the necessary resources to each application;

– a level of criticality (descending from A to E) is assigned to each function. The separation of the functions is performed by spatial isolation in the memory and temporal separation of the module processor. It is forbidden for a function to exchange data with a function of higher criticality (memory);

– certification is incremental: the module is certified automatically if the core and each application are certified individually.

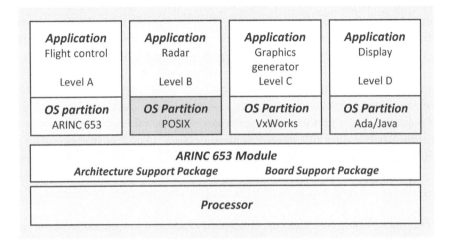

Figure 2.4. *Structure of an IMA module*

2.2. Data transmission

The digital transmission of network data shall be provided with essential minimum properties when implemented in critical safety systems:

– the *events determinism*, which guarantees that no data is lost;

– the *temporal determinism*, which ensures that the data is transmitted with a delay (or latency) bounded and showing low fluctuations (jitter);

– the *integrity*, which ensures that data is transmitted without error.

To meet these needs, the solutions implemented must in particular manage the collisions on the databus (simultaneous requests for data transmission) and detect transmission errors. Data is transmitted in digital form, in series (one after the other, that is to say, bit after bit) of structured blocks of binary data that form the frame. Transmission lines are generally in the form of twisted pair(s) of wire(s), possibly shielded. The shielded quadrax cables comprise two twisted pairs. Except for ARINC 629[1], the differential voltage level between the twisted pair wires defines the logic state (0 or 1) of the binary information transmitted. This allows better immunity against disturbances as they affect the two conductors in the same way. The shield reinforces this immunity by limiting the magnitude of the disturbance picked up by the two wires of the twisted pair. Differential voltage levels are typically a few volts. They can go down to ±300 mV with Low Voltage Differential Signaling (LVDS), which improves the transmission speed while reducing power consumption.

In one-way communication, for example, data can only flow in one direction along the transmission line (*simplex* if the direction is fixed and *half duplex* if the direction can be reversed in time). When data move in both directions, it is called *full duplex* communication. Data transmission can be asynchronous (the data is transmitted as soon as possible) or synchronous (the data is transmitted at specific times). There is talk of isochronous transmission when the time interval between these moments is constant.

The main topologies that can be implemented and combined to achieve a data network are shown in Figure 2.5. Elements can be connected in different topologies: as a series (daisy chain)/tree/loop/star and in the form of a bus. In a *star* topology, all stations are connected to a central node. In unidirectional star architectures, only the central node can emit. In a *bus* topology, all stations are connected to a common bus. In *multi-drop* bus (or bus-drop), the information flows only in one direction along the bus (simplex) from a single transmitting station (*transmitter*) to other stations that are only receiving (*receiver*). The term *multi-point* bus is applied when information can flow in both directions on the bus (full-duplex)[2], where each station can transmit or receive (*transceiver*).

Table 2.1 shows, synthetically, the principles and characteristics of the most used digital communication standards. Concepts and architectures are

---

1 For this standard, it is the level of current flowing which defines the logic state.
2 However, only one direction is active at a given time.

briefly discussed in the following sections. For more details, the reader may consult such references as [GWA 06, MOI 13, OBE 12, SCH 08].

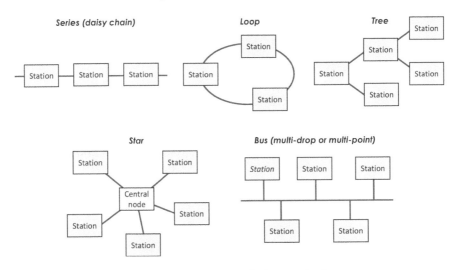

Figure 2.5. *Data transmission topologies*

2.2.1. *CAN*

The CAN (Controller Area Network) standard was originally developed for the automotive industry, in order to reduce the number of point-to-point connections. In aviation, the CAN bus communication is, for example, applied to exchange information relative to the actions of the pilots between two active side-sticks.

The CAN standard defines a digital communication series, with a high integrity, between microcontrollers and components, according to a multi-point architecture [RAO 09]. A station can begin transmitting when there is no activity on the bus (Carrier Sense Multiple Access or CSMA). Each station can send a frame which is then sent to all stations (including itself). The frame has no source or destination address, but an identifier allows each station to filter any relevant information. The identifier includes a priority index that avoids collisions when attempting simultaneous broadcast by several stations: arbitration is conceded by only allocating priority to the station originating the frame (Arbitration on Message Priority or AMP), such that other stations then give way to the bus/access of the originating station. The other frames can then only be transmitted following a new request, once the bus is inactive again. With this approach, communication between stations is event deterministic, asynchronous but without a delay guarantee. The high integrity is obtained by multiple error detections carried out by each station, at the level of each bit and at the level of the total frame.

2.2.2. RS422 and RS485

The RS422 and RS485 standards [SOL 10] define only the electrical characteristics of the digital transmission and the transmission protocol is not imposed. The RS422 standard only allows for a multi-drop architecture with one transmitter and up to 10 receivers. The RS485 standard allows for it to achieve a multi-point architecture with up to 32 stations.

2.2.3. ARINC 429 [AIM 10a]

The ARINC 429 standard was developed to facilitate interoperability and interchangeability of avionics when present in the form of LRU elements. It began to be implemented on commercial airplanes with the Boeing 757, the Airbus A310, and the MD-11 McDonnell. The ARINC 429 bus is an asynchronous bus. It is also unidirectional: a single data *transmitter* (or Tx) can be connected to a maximum of 20 data *receivers* (or Rx). The ARINC 429 is known for its protocol simplicity and reliability. As shown in Figure 2.6, it is necessary to install as many communications lines, with as many elements that have to operate as a transmitter (including collating the receipt of data). Therefore, the ARINC 429 standard is less efficient in wiring for complex systems where there are many elements and many interactions between elements.

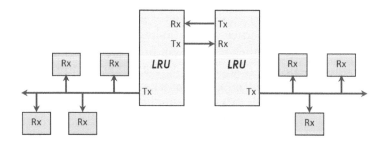

Figure 2.6. *Example topology allowed by the ARINC 429 standard*

	CAN [ISO 11519/11898]	RS422	RS485	ARINC 429	Mil-STD 1553B	ARINC 629	AFDX ARINC 664	IEEE1394b SAE 5643	TPP/C
Date	1994 (low speed) 2003 (high speed)	1994	1998	1974	1978	1995	1997	1995/1998	2002
Protocol	According to variants			Not applicable	Command/ response	DTSA			
Transmission Class	Multipoint	Multidrop	Multipoint	Single source/ multiple sinks	Multiple source/ multiple sinks	Multiple source/ multiple sinks			Time triggered
Maximum Frame Length	128 bit	5–8 bit (data)	5–8 bit (data)	32 bit	20 bit	20 bit	1,518 bytes		240 bytes
Data length	0–64 bit			19 bit		16 bit			
Maximum Number of Stations	32 to minimize delays	1T 10R	32TR	1T 20R	1 bus controller 31 remote terminals	120TR	64,000 (virtual link)		64
Maximum Transmission Speed	1 Mbit/s 125 kb/s	10 Mbit/s	10 Mbit/s	100 kbit/s	1Mbit/s	2 Mbit/s	10/100 Mbit/s		25 Mbit/s
Maximum Bus Length	40 m 500 m	1,200 m	1,200 m	65 m/ 150 m		180 m	<100 m		

T = transmitter, R = receiver, TR = transceiver

Table 2.1. *Main digital data transmission standards used for actuation*

2.2.4. MIL-STD-1553B [AIM 10b]

The MIL-STD-1553B standard is an evolution of the 1553A (1973) standard for military applications. The first applications appeared on the General Dynamics F-16 fighter and on the Hugues AH-64 helicopter. The standard uses a bi-directional dual-redundant bus with a centralized bus control using a command/response type protocol: a single *bus controller* which manages the exchanges between stations (*remote terminals*), of which a maximum of 31 can be connected to the bus. No station can transmit data without being invited to do so by the controller. To be physically redundant, the bus uses two communication lines (see Figure 2.7). In case of damage to the primary bus, the secondary bus can take over to provide a fail-operative redundancy. The failure node that accounts for the controller is removed by the presence of a second controller with hot redundancy that can be activated as a backup.

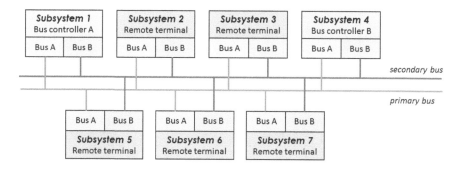

Figure 2.7. *Example of multi-point topology by the MIL-STD-1553B redundant bus. For a color version of this figure, see www.iste.co.uk/mare/aerospace2.zip*

2.2.5. ARINC 629

The ARINC 629 standard was developed by Boeing to increase flexibility, speed of transmission and immunity to Electro-Magnetic Interference (EMI) while reducing wiring complexity. This standard is used on the Boeing B777 in a triplex version for the flight controls that involve remote electronic actuator controls (Figure 1.19).

The ARINC 629 implements a bi-directional and dual-redundant bus (or even triple or quadruple): it can interconnect up to 128 stations operating in transmission or reception. Control of the bus is distributed: each station independently manages its bus access based on the activity of the bus, using a protocol based on the Dynamic Time Slot Allocation (DTSA). In contrast to other standards, logic value is defined by the current (not voltage) circulating on the twisted pair wires, which limits the effect of EMI, even in the absence of shielding.

2.2.6. AS-5643/IEEE-1394b

The combination of the AS-5643 and IEEE-1394b standards allows for low-cost, synchronous data transmission networks, which are robust and deterministic and hence suited for aeronautical constraints. It is used, for example, on the Lockheed Martin F-35 combat aircraft.

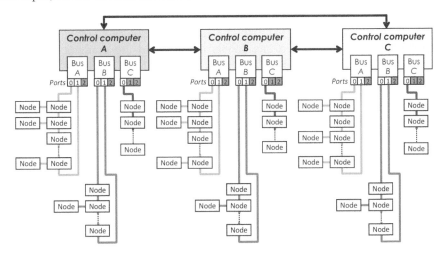

Figure 2.8. *Triplex architecture to the standard AS-5643/IEEE-1694b, representative of the implementation adopted by the F35 Lightening (from [BAI 07]). For a color version of this figure, see www.iste.co.uk/mare/aerospace2.zip*

The IEEE-1394b standard [INS 02], also called *Firewire 2*, originates from mainstream computing. The associated electronics has excellent maturity; it is available Commercially Off-The-Shelf (COTS) at a low cost and all types of architectures can be implemented. The AS-5643 standard [SOC 04] benefits from the IEEE-1394b standard by combining a Vehicle Management Computer (VMC) and remote nodes (LRU). It authorizes the voltage amplification at the nodes, allowing a 500 Mbit/s transmission rate on distances greater than 15 m. It defines the protocol in which the command computer triggers the Start Of Frame (STOF), which typically operates at a frequency of 80 Hz. Messages are sent to all the remote nodes and taken into account only if they have subscribed to this message (Anonymous Messaging Subscriber or ASM). The transmitted data contains a temporal offset that defines the data transmission delay with reference to STOF. Thus, the relevant remote node can be prepared to receive data at a specific time. This protocol ensures network determinism. Figure 2.8 shows a redundant network architecture based on the example of the AS-5643 and IEEE-1394b standards. It is noted that series, tree and/or loop topologies can be combined.

2.2.7. AFDX (ARINC 664 Part 7)

The AFDX protocol (Avionics Full-DupleX Ethernet switching) was developed by Airbus in the late 1990s to facilitate the application of the concept of integrated modular avionics [BUT 07]. Implemented on the Airbus A380/400M/350, the Boeing B787, the Comac C919/ARJ21 and the Sukhoi Superjet 100, it is based on the Ethernet standard (or IEEE 802.3), which benefits from a great maturity since its introduction to commercial products in the early 1970s. AFDX introduces some adaptations to this standard to meet the needs of data exchange in avionics applications: speed, reliability and determinism.

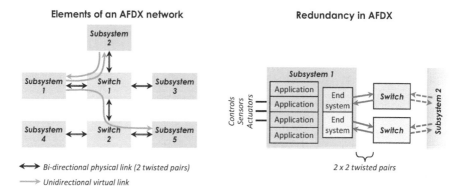

Figure 2.9. *Topology and redundancy for the AFDX bus. For a color version of this figure, see www.iste.co.uk/mare/aerospace2.zip*

As shown on the left-hand side of Figure 2.9, an AFDX network consists of three types of elements: system terminal users (*end-system*) contained in the stations (*sub-system*) which interface the applications that communicate through the AFDX network, the *switches* and (physical or virtual) links. The AFDX is made redundant by duplicating switches and links (shown in the right side of Figure 2.9).

1) Switches:

Unfortunately, an Ethernet network is not deterministic: when two stations want to transmit simultaneously, a collision occurs on the bus and the attempted transmission must be renewed without any guarantee as to the period of time for success. In the AFDX, successful transmission is guaranteed in a limited time period by the switches which interconnect the stations. The switches *filter* and *police* the traffic between subsystems (*traffic filtering* and *policing*). They only allow for traffic that is predefined by virtual links and protect against inconsistent transmissions (*babbling idiot*) or erroneous data packets. They guarantee for the minimum bandwidth and the maximum latencies as defined by the configuration of the network. A switch can be connected with up to 24 stations.

2) Links:

The transmission on the physical links is bidirectional. Each link is therefore comprised of two twisted pairs (one for each direction of information transmission). The stations interface with twisted pairs via user terminals. The AFDX uses the concept of virtual link. Each virtual link describes a logical, unidirectional connection between a station and the other stations. In total, 2^{16} virtual links can be defined. In each station there is *traffic shaping*: it defines the authorized recipients (the virtual links) and the bandwidth contract allocated to it, for the transmission of data to each of them.

3) Terminal users:

Subsystems manage their own *traffic reception*. Each end-system checks the integrity of the frames (detection and elimination of invalid frames) arriving on the redundant buses. The first frame recognized as valid is transmitted to the receiving application it hosts. The redundant frames are eliminated.

The previous concepts (virtual link, traffic shaping, traffic policing and traffic reception) make the AFDX a bi-directional network, which has redundant, distributed control and (pseudo-) deterministic properties. High speed and flexibility come from the dissociation of physical and virtual links: a virtual link can travel by different physical links as per their availability, only the exchanges defined by the virtual links being allowed. The micro-AFDX is a simplified version of the AFDX, which is well suited for the transmission of actuation data.

2.2.8. *Triggered time protocol (TTP/C)*

The TTP/C protocol is designed to transmit data with strong real-time constraints, at a low cost and with fault tolerance. It is for example, implemented on the Embraer Legacy 450 business aircraft and Bombardier C series commercial aircraft. To avoid collisions, the protocol combines:

– a strategy of Time Division Multiple Access (TDMA) which allocates to each station a fixed time window, predetermined by the dispatch table, in which the station is authorized to issue. In a TDMA *round*, each station is authorized in turn to issue once. The issue is then carried to all stations;

– a dispatch table (Message DEscriptor List or MDEL), predefined and known to all stations, indicates which station can transmit messages and when;

– a distributed time synchronization.

Stations can be interconnected according to a two-channel architecture, bus or star type [ELM 03]: communication is thus ensured even in the case of failure of one channel (single fail-operative). The left side of Figure 2.10 shows the basic TTP/C node (Smallest Replaceable Unit or SRU) corresponding to a multi-point topology. The protocol is implemented in a bus controller. Each controller has a *bus guardian* which ensures that the protocol processor sends messages according to the dispatch table. The guardian also ensures that the elements do not transmit if they do not work properly *(fail-silent)*. In the star topology, the two bus guardians are centralized, as shown on the right-hand side of Figure 2.10.

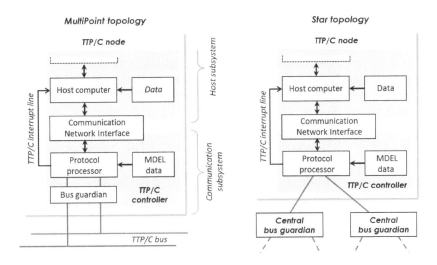

Figure 2.10. *Multi-point or star topologies for standard Time Triggered Protocol. For a color version of this figure, see www.iste.co.uk/mare/aerospace2.zip*

2.3. Evolutions in data transmission

The previous sections have shown how the evolution of SbW (partial and complete, analog and digital, federated and modular) has gradually facilitated geometric integration, to increase the performances of existing functions and/or enabled new functions to be implemented. The increase in data rates, scalability and simplifications in wiring can be further improved by reducing the number of electric wires[3]. For this, it is possible to use other means for transmitting information: a single cable for transmitting power and information, either via fiber optics or wireless transmission. In all cases, there are strong constraints regarding reliability (especially detection and isolation of transmission errors) and immunity to electromagnetic interference, High Intensity Radiated Fields (HIRF) and to Electro Magnetic Pulses (EMP) from lightning or a nuclear explosion. The flight control actuators are particularly affected by these interferences because they do not benefit from the Faraday cage/shield effect produced by the aircraft's structure, particularly in the landing or take-off phases, when the slats and flaps are deployed.

---

3 [SMI 07] highlights, however, that while this reduction eases integration and reconfiguration, the past experience has shown it to increase the price of LRUs and in actual fact does not necessarily reduce the overall mass and/or complexity of aircraft.

2.3.1. *Power over data and power line communication*

The transmission of data over power lines (Power Line Communications or PLC) or power transmission over data networks (Power over Data or POD) can reduce the amount of electrical wires.

In the United States, PLC technology for aircrafts was evaluated in the mid-2000s for the transmission of bi-directional measurement or clock signals over 28 V DC power lines [JON 06, TIA 10]. The choice to return current via the metal frame of the aircraft thus negatively impacted on latency and the data rate. Later, a study evaluated the interest and feasibility of the PLC for flight controls [OBR 08]. The typical application was to use the PLC in an IMA architecture for transmitting data between the flight control computer and the remote electronics. It was evaluated for a long-haul, wide-body, passenger plane. The PLC was deployed on six independent networks in total linking six computers to 27 actuators, with a latency requirement of 1 ms for distances of 37–68 m. The voltage level to be superimposed on the DC voltage, in order to transmit data, was limited to ±2 V for it to be considered as a ripple, according to standards, with regards to the elements connected to the DC line. For a network comprising five remote electronics, the estimated transmission rate was 2 Mbit/s, the same as the ARINC 629 standard. In conclusion, the study showed that the PLC allowed for the removal of more than 3 km of wires, translating to a mass gain in excess of 17 kg, if it were instead implemented as the backup channel. It also highlighted the lack of off-the-shelf components and barriers to be overcome in terms of the presence of a switch on the DC line. The European project TAUPE (Transmissions in Aircraft on Unique Path wirEs), which took place between September 2008 and February 2012, demonstrated, under laboratory conditions, the feasibility of simultaneous data and power transmission over a single electrical cable for non-critical applications. POD and PLC technologies were combined [DÉG 10] and applied to the cockpit displays and for the cabin lighting. The *Technology Readiness Level* was rated at TRL4[4]. For an Airbus A380, the maximum potential gain was estimated at 36 km of cables (from a total of about 500 km), translating as a

---

4 The TRL scale, initially implemented by US government agencies, varies from Level 1 (the basic concepts) to Level 9 (the product is operational). It is often assumed that beyond Level 6, there is a shift from development to industrialization. At TRL4, the technological implementation of the concept is validated in a laboratory environment. For TRL5 and TRL6, it is validated and demonstrated in a realistic environment. The example of the PbW actuators shows that it takes sometimes two decades to go from TRL4 to TRL9.

gain in mass of 360 kg. The European MPbus (Modular Power bus), launched in April 2014, relates more to space applications. It aims to reduce the mass of the DC power bus of a satellite, in particular by using it to transmit data. Other studies have also been conducted on the onboard telephony and Internet access of aircrafts, and based on PLC broadband technology [DAM 08]. Recent work [MOR 16] also showed that the POD or PLC concepts can be achieved by using optical fiber transmission.

Ultimately, it appears that the technological maturity of the POD and PLC concepts is far from sufficient for it to be applicable to critical safety functions such as flight controls.

2.3.2. *Optical data transmission (Signal-by-Light or SbL)*

The use of optical fibers for the transmission of data has many advantages. This is shown in Table 2.2 which compares optical fibers used in commercial aircrafts to its electrical equivalent, within the same operating temperature range. In this example, it appears that optical fibers produce 16.6 times less attenuation over 100 m. The advantages for integration are equally impressive: 8.7 times lighter, taking 2.2 times less space in diameter and with a bending radius of less than 15%.

	Fiber optic	Shielded twisted pair
Reference manufacturer	ABS0963-003LF Nexan	STUDY 133026 Nexan
Composition	Silicone fiber	Plated copper alloy, 2 × 19 strands
Active diameter	62/125 μm	0.12 mm
External diameter	1.8 mm	4.1 mm
Linear density	<4 kg/km	35 kg/km
Maximum bending radius	20 mm	23 mm
Attenuation	4 dB.km for 850 nm 2 dB.km for 1,350 nm	6.6 dB for 100 m
Bandwidth	400 MHz.km for 850 nm 1,000 MHz.km for 1,350 nm	Unspecified
Operating temperature	−55 to 125°C	−55 to 125°C

Table 2.2. *Comparison between optical fiber and shielded twisted pair wiring*

Optical fibers cannot cause a short circuit, which limits the risk of fire and fault propagation. It is impervious to EMI, HIRF and EMP, which makes it even more interesting in the application of carbon fiber reinforced structures for aircrafts, which otherwise impose severe constraints in terms of the shielding and continuity of wired connections. However, the main disadvantages of fiber optics lie in its implementation, particularly for connections that require it to have a very high level of cleanliness in order to limit losses.

Table 2.3 illustrates the efforts, over four decades, which aims to demonstrate and establish the maturity for the application of fiber optics to transmit flight control data. Early works contended to remove all wire transmission to/from the actuator in favor of optical fibers (fully integrated FbL) with the view to improve immunity against electromagnetic disturbances. High power lasers were found necessary to drive the moving element of servovalves and mode solenoid valves. Finally, this gradually led to the acceptance that the power required for the actuator control function needs to be supplied electrically. Moreover, the recent commercial aircraft programs implementing electrically signaled smart actuators, suggest that existing state-of-the-art wiring technology can now meet the requirements related to the electromagnetic environment through Signal-by-Wire transmissions to/from the actuators. This being said, optical fibers are still the preferred medium to transmit data, especially for high data transfer rates between remote elements. For example, 1.7 km of optical fibers are used on the Boeing 787 to carry 110 optical data links. On the Airbus A380, 171 data links are implemented through 2.4 km of optical fibers, four times more than the Airbus A330/340 [BOU 11].

Year	*Manufacturer model*	*Aircraft type and main results*
1976	Boeing YC-14	Short takeoff and landing for transport aircraft Optical data transmission
1982	Vough A-7D Corsair	Carrier-borne fighter Test flight with flight controls to simplex optical signals

Year	Aircraft	Description
1980	NorthAmerican T-2C	Training aircraft. HOFCAS/AFCAS project [KOH 80] Electrohydraulic yaw actuator to optical transmission of signals and power metering by DDV Electrical power produced internally to the actuator
1983	McDonnell Harrier AV-8B	Vertical takeoff fighter Optical data transmission implemented into production
1987	Sikorsky UH-60A Blackhawk	Military helicopter. ADOCS project [SHE 85, TER 89] Flight tests for optoelectronics flight controls with optical position sensors (mini side-stick, pedals and primary cylinders) and optical connections between the sensors, the computer (AFCS and SCAS) and the actuators
1988	Bölkow BO105-S3	Light helicopter. OPST project [FAU 93, STO 89] Yaw control test flights for optoelectronic signaled triplex gyro and position sensors for optical output
1993	BAC1-11	Commercial aircraft Flight test for the optical link between spoiler computer and two smart spoiler actuators
1994–1997	McDonnell Douglas F/A-18 Hornet + MD90	FLASH project [HAL 95a, HAL 95b, HAR 95, TOD 96, ZAV 97] Maturation of connectors, installation and maintenance [HAL 95a] two stage servovalve to single optical input [HAR 95] The active side-stick to optical output MD90 demonstration flight: EMA aileron trim to optical signals [TOD 98]
2002	Eurocopter EC 135	Light helicopter ACT/FHS [BIC 03] Flight control flight tests for optoelectronic 4-axes triplex Electrohydraulic smart actuators
2008	Gulfstream G650	Business jet Flight tests of optically signaled spoilers

Table 2.3. *Some striking examples of SbL to the flight controls*

2.3.3. *Wireless data transmission (Signal-by-WireLess or SbWL[5])*

The removal of electric cables in favor of radio transmission is attractive to many aspects concerning the different life phases of an aircraft [ELG 10]:

– in the development phase of an aircraft, it allows for the removal of routing-related activities and does not induce any change due to the design constraints;

– in the integration phase, it reduces the activities linked to production of harnesses, integration in the airframe[6] and the time taken for testing;

– during the operational life of the aircraft, it eliminates the risks associated with aging harnesses and facilitates the detection of defects;

– in total, according to [GRA 09], it has the potential to reduce the specific consumption of an airliner by 12%.

However, SbWL introduces strong constraints vis-à-vis electromagnetic compatibility, as well as safety and compliance with certification rules. In terms of flight controls, the principle of FbWL was demonstrated by flights in September 2008. It concerned the operation of a spoiler on the G650 Gulfstream test aircraft. As it stands, Hertzian transmission remains confined to certain non-critical applications: in-flight entertainment, smoke detectors, condition monitoring, etc. The Boeing 787 uses FbWL for the emergency lighting system (Wireless Emergency Lighting System or WELS) which is activated to evacuate the cabin in the event of an emergency.

Finally, the FbWL appears, for now, to exist as a dissimilar solution to other means of transmitting information to the flight controls. As such, it could be used as a backup channel, for example in the case of structural damage caused by engine rotor burst.

---

5 In related literature it is also referred to as FbWL and sometimes as FbWLSS.
6 According to [ELG 10], the overall cost associated with the installation of wiring is 2,200 €/kg.

3
Power-by-Wire

This chapter focuses on the power view for electrically supplied actuation. In the discussion to come, the qualifier "conventional" will be used to refer to actuators that are hydraulically supplied, that is to say powered at a constant pressure via a centralized hydraulic power network throughout the aircraft. This type of actuator has been studied in detail in Volume 1, and the reader is invited to make reference to this work for greater in-depth explanations and additional details.

The organization of this chapter follows a pattern of thought most frequently encountered when looking at potential advancements. Upon initial investigation, we could say that it, is natural to consider exclusively the disadvantages of the existing solutions to be replaced, whilst focusing on the inherent benefits of the proposed new solutions. Only upon closer inspection can the intrinsic benefits to conventional solutions be identified and given importance, and the new disadvantages (or challenges) of the proposed solutions be brought into question. Figure 1.5 of Volume 1 shows the major power functions required by actuation: the generation, transport and distribution, and metering of power. Power architectures and hydraulic technology are well established and have been applied for many decades for all types of aircrafts. This level of maturity, regarding these major power functions, has allowed for a good compromise in their application: a balance between power density, reliability and dynamic performances. However, the rapid improvement of technology maturity in the field of electronic solutions for actuators makes for compelling arguments to reduce or eliminate the inherent drawbacks associated with the use of hydraulics.

3.1. Disadvantages of hydraulic power transmission

Figure 3.1 uses the example of the Airbus A330 to introduce, in a simplified form, a conventional 3H power architecture. In this type of architecture, hydraulic power is distributed by three independent power networks (referred to as green, yellow and blue). For each network in this configuration, power is generated centrally: four main pumps are driven by each of the two engines (Engine Driven Pump (EDP)) and one pump for each network is driven by an electric motor (Electro Mechanical Pump (EMP)). As an ultimate backup, a pump may be driven by dynamic air through a turbine (Ram Air Turbine (RAT)). The assignment of power users to a specific network is driven by reliability requirements. Thus, the most critical functions such as flight controls can be carried out in case of failure of two out of the three hydraulic power systems.

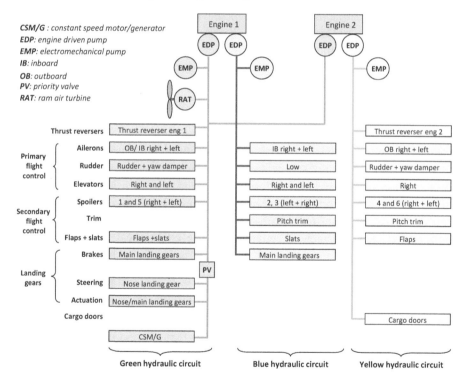

Figure 3.1. *An example of conventional architecture for centralized 3H generation as per the Airbus A330. For a color version of this figure, see www.iste.co.uk/mare/aerospace2.zip*

The following sections outline the specific disadvantages regarding this type of architecture.

3.1.1. *Power capacity of hydraulic pumps*

Pumps have the function to supply fluid at a constant pressure (usually 210 bar, and 350 bar on more recent aircraft), regardless of the flow rate requested by the users. These are predominantly driven by mechanical power which is drawn from the engines of the aircraft, as per a constant kinematic ratio. Their speed of rotation is directly proportional to that of their drive which itself depends on the phase of flight. The maximum flow delivered by the pumps is also functionally proportional to their speed of rotation. This direct dependence on engine speed is disadvantageous: the pumps must be sized to cover all the possible flow requirements, even for the lowest drive speeds (for example, when the engine is idle during approach or descent). The main pumps of a single-aisle aircraft, for example, deliver for these conditions less than 60% of the maximum available flow at takeoff.

3.1.2. *Hydraulic pump efficiency*

For a given actuator, the level of supply pressure defines the maximum drive force that the actuator can generate to the load at stall. However, it appears that on a typical flight mission (Figure 1.4 of Volume 1), the average force and speed needs only represent some tens of percent of their maximum value. Hydraulic pumps therefore consistently deliver fluid, at the chosen supply pressure, working predominantly at nominal speed (long-haul aircraft) and deliver, on average, a low flow in respect to their rated capacity. Under these conditions, the fluid flow rate delivered by the pump is no longer sufficient to evacuate the heat, which is produced by energy losses within the pump (leakage, friction). Therefore, in order to keep conditions cool, it is necessary to establish a permanent flow inside the pump, from the suction inlet to the tank through the drain circuit; a process that greatly penalizes the average efficiency of hydraulic power generation. In the example of the main pump in a single-aisle aircraft, drainage permanently consumes over 10% of the rated power of the pump drive. On top of this, it is essential to consider the permanent internal leakage within the equipment supplied by the pump, and which for its own part needs a permanent supply of 8% the nominal flow of the pump. Thus, without even accounting for the flow demand for actuation, the pump's draw on the drive shaft is between 15 and 20% of the rated hydraulic power that the pump can deliver.

3.1.3. *Centralized power generation*

Hydraulic power is generated, in normal mode, by the engines of the aircraft and in turn sent to numerous actuators, some of which may be located several tens of meters away. The fluid-filled hydraulic lines, their attachments to the airframe, the connections and the fluid conditioning functions incur a strong mass penalty for the conventional solution. On larger aircrafts, it is estimated that 75% of the mass for the hydraulic system is accounted for by the power distribution network, with only 25% accounting for the actual equipment (pumps and actuators). Focusing on compromises between mass and energy dissipation in power networks thus has an impact on both the power density and the efficiency of power distribution.

3.1.4. *Power transmission by mass transfer*

In hydraulics, the power transmission function requires the physical transfer of material: at the pumps the hydraulic fluid is pressurized, providing it with hydrostatic energy. This is then put in motion at low speed (a few m/s) to transfer this energy to the actuators. This transportation of physical material is a source of numerous disadvantages.

3.1.4.1. *Negative impacts on the environment and risks to people*

On the one hand, non-commercial aircrafts often use a synthetic, mineral-based, hydraulic fluid, which is not environmentally friendly. On the other hand, the fluid used by commercial aircrafts is far less flammable, but is transformed into an acid in the presence of water. As such it is far more aggressive to people and the environment.

3.1.4.2. *Fluids conditioning needs*

The transferred power, produced by the physical transport of material, requires that the energy vector be properly conditioned to fulfill its main function (the transmission of hydrostatic energy) as well as its secondary functions (lubrication, heat conveyor). The hydraulic fluid must be present in sufficient quantity, without any solid/liquid/gas pollutants. It must expand or retract freely and it must be maintained in its operational temperature range. This imposes severe constraints on the design, production and operation of hydraulic systems: cleaning and maintenance, pressurization, purge, contamination controls, cooling/heating, etc.

3.1.4.3. Difficult reconfiguration

To satisfy the constraints of reliability, multiple hydraulic power networks are installed. The constraints of separation and independence prohibit the interconnection of hydraulics between these networks, so as to avoid the spread of leaks or pollution. This deprives the designers of opportunities to reconfigure or make redundant the power pathways to actuators. On military aircrafts, this particular constraint increases their vulnerability to enemy fire.

3.1.4.4. Aircraft integration constraints

The integration of hydraulic power distribution systems in the aircraft is constraining. At the level of routing, there must be topological segregation that avoids critical areas (fire zone, passenger cabin) and limits bend radius to avoid stress concentration within the tubing.

3.1.5. Control of power by energy dissipation

The power metering function is performed by control of a valve opening: the part of the supply pressure that is not required by the actuated load is spent; this is due to the throttling losses at the metering level according to its opening. Power consumption is thus independent of force production, due to the fact that all the supply pressure is used for the load and crossing the variable orifices of the power valve. As a result, the power consumption is only determined by the speed of the load, that is to say, by the flow drawn by the power network. The efficiency of power metering is thus mainly dependent on the relation between the differential pressure required by the load and the net supply pressure. As the average force required by the load to operate is on average low with respect to the maximum force available, the average efficiency of power metering is low.

The metering function can also cause other power losses. On a conventional SbW actuator, the metering function is performed by an electrohydraulic servovalve[1]. This servovalve uses a flow-dividing hydraulic pilot stage to generate sufficient force on the mobile element of the hydraulic power valve. This amplification stage generates a permanent consumption of pressurized fluid (for example 0.2 l/min for each servovalve). Thus, for 35 flight control actuators supplied at 207 bar and with a return pressure of 5 bar, the constant leakage of the servovalves' pilot stages creates a permanent power loss of 2.3 kW.

Finally, it is also worth mentioning that the throttling loss in a hydraulic resistor does not allow for the recovery of energy when the load to be moved operates in aiding load power quadrant.

3.2. Electrical power versus hydraulic power

Table 3.1 synthetically compares the advantages and disadvantages of hydraulic versus electric technologies for actuation. Further detail and explanation for the electrical part will be given in later chapters.

Merits	Hydraulic	Electric
Power density at the level of the actuators	Excellent	Average
Power density at the level of the power network	Poor (tubes, fasteners, fluid)	Good (3-phase)
Efficiency of control of power	Generally low (throttling losses)	Excellent (power-on-demand)
Direct mechanical power transmission to the load (direct-drive)	Easy with linear actuators	Mechanical reducer generally mandatory
Evacuation of the heat generated by energy losses in the actuator	Excellent, due to the hydraulic fluid returning to the reservoir	Poor, local exchange with the ambient
Power management functions (clutch, brake, damping, force limitation)	Easy, compact, lightweight and efficient in the hydraulic field	Often inefficient if applied to the field of electrics. Heavy or bulky in the mechanical field
Inertia reflected by the actuator	Low	High

---

1 See Volume 1, section 5.5.

Options and ease of command	Average	Excellent
Dynamic reconfiguration of power paths (redundancy)	Difficult	Easy
Integration and operation constraints (segregation, installation, maintenance)	Strong (fire, leaks, pollution, bleeding)	Moderate (EMI, HIRF) Built-in test options
Technology maturity level	Excellent	Low return of operational experience
Potential evolution	Moderate to weak	Strong
Environment and human friendliness	Poor	Good

Table 3.1. *Comparison of hydraulic and electrical power technology*

From a simplistic perspective, hydraulic technology has three major disadvantages:

– energy efficiency: the average efficiency of power generation and control through throttling is poor;

– power networks: the hydraulic power networks are heavy and impose strong constraints at all levels (design, production, operation);

– environment: the hydraulic fluid is aggressive for people and the natural environment.

It turns out that these are precisely the aspects in which electric technology has its best qualities. The evolution of power actuation to becoming more, if not completely, electrical aims to take advantage of these qualities. However, as very often is the case for advancements in aerospace, this change occurs incrementally at both the level of the actuator and at the level of the network or aircraft: the improvement in hydraulic technology, the combination of both hydraulic and electric technologies, then the removal of hydraulic technology. The main concepts generated by this evolution are summarized in Figure 3.2, to be read from top to bottom, from fully hydraulic to full electrical, respectively.

All these developments can be identified through a comparison of the power architectures of the Airbus A380 and Boeing B787 (Figures 3.3 and 3.4, respectively) with the all-hydraulic architecture in Figure 3.1.

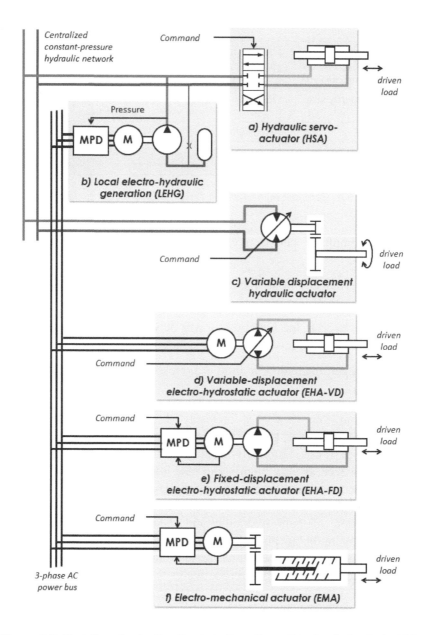

Figure 3.2. *Evolution of actuation (top to bottom, from all hydraulic to all electric). For a color version of this figure, see www.iste.co.uk/mare/aerospace2.zip*

Power-by-Wire 67

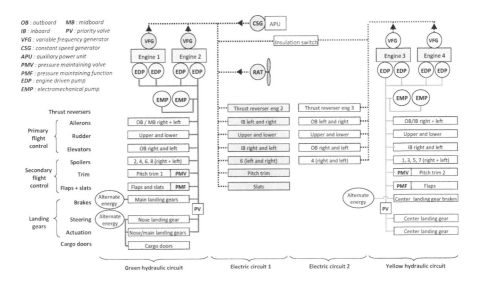

Figure 3.3. *Type 2H–2E power architecture for actuation functions on the Airbus A380 [MAR 04]. For a color version of this figure, see www.iste.co.uk/mare/aerospace2.zip*

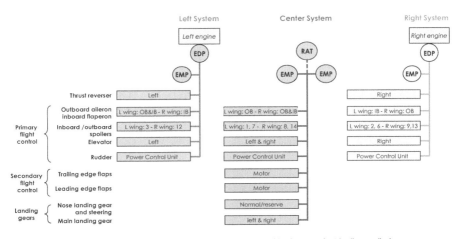

Figure 3.4. *Power architecture for actuation functions on the Boeing B787. For a color version of this figure, see www.iste.co.uk/mare/aerospace2.zip*

For both of these "more electric" commercial aircraft, the main changes regarding actuation are listed in Table 3.2.

	Airbus A380 (EIS 2007)	Boeing B787 (EIS 2011)
Improving hydraulically powered solutions	Operating pressure increased by 67% Displacement control for high-lift actuators	Operating pressure increased by 67%
Combination of hydraulic and electric	Electro-hydrostatic actuators for primary flight controls (backup) Electrically supplied local hydraulic power generation (backup)	
Removal of hydraulic technology (*hydraulic-less*)	Electro-mechanical actuator for the backup channel of the trim horizontal stabilizer Electric thrust reversers The complete removal of one hydraulic system, as compared to conventional solutions	Electro-mechanical actuator of the trim horizontal stabilizer Electric brakes Electro-mechanical spoiler actuators (4 of 14)

Table 3.2. *More or fully electrical actuation on the Airbus A380 and Boeing B787*

3.3. Improving hydraulically supplied solutions

For power transmission, there are potentially several ways to improve hydraulics, without resorting to electrical transmission. Most of the solutions have been designed to reduce energy losses and increase the power density of distribution networks.

3.3.1. *Reduction of energy losses in actuators*

3.3.1.1. *Reduction of servovalve leakages*

Electrohydraulic servovalves were presented in section 5.5 of Volume 1. Their internal leakage has two origins: leaks at null opening for the power stage and permanent leakage at the pilot stage. The leaks at null opening can be reduced by increasing valve overlap. In practice, this solution is seldom used because it generates a dead zone effect on the command, altering the sensitivity of the actuator to small amplitude commands. The permanent leakage at the pilot stage may itself be eliminated by direct electromechanical piloting of the power stage, which gave rise to the Direct Drive Valve (DDV) (see Volume 1, section 5.5.5). These are found on some military aircraft and helicopters such as the Northrop B2, the Saab JAS 39, the Dassault Rafale, the Eurofighter and the NH Industries NH90.

3.3.1.2. *Displacement control at destination*

This solution was presented in section 5.2.2 of Volume 1, as was the principle behind it, shown in Figure 3.2(c). Displacement adjustment can be applied to actuators employing a hydraulic motor, typically for operating high lift devices (slats and flaps). The control of power can then be carried out by acting on the displacement of the hydraulic motor, the latter being fed with constant pressure by the hydraulic network. Functionally, only the power required by the load is taken up by the hydraulic source (power on demand). This was introduced by Airbus for operating the slats and flaps on the A380 [BOW 04] and then spread to following programs (A350). This development is shown in Figure 3.5, which compares the power architectures used for actuation of the slats and flaps of the A330 and A380.

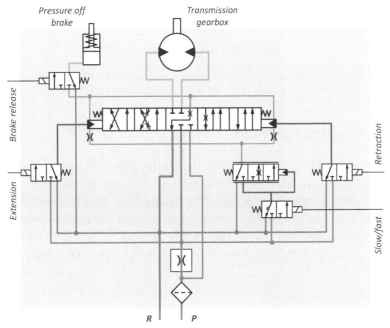

a) Actuation of slats on the Airbus A330

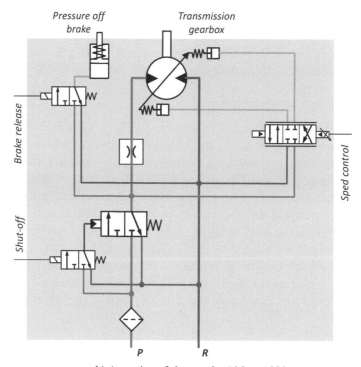

b) Actuation of slats on the Airbus A380

Figure 3.5. *From valve control to displacement control for slats actuation. For a color version of this figure, see www.iste.co.uk/mare/aerospace2.zip*

3.3.2. *Increased network power density*

The power density in hydraulic systems can be improved by increasing the operating pressure (see Volume 1, section 1.3.3). The gain is mainly due to the reduced volume of fluid in the pipes. After having been used for more than 10 years in military aircrafts, this has been put to practice on recent commercial aircraft programs (Airbus A380 and A350, Boeing B787). The operating pressure was increased by 67% from 207 to 345 bar (3,000–5,000 psi).

3.3.3. *Other concepts*

Several concepts have been investigated and published, for example in [GRE 88], but they may not have been put into aerospace production.

Despite this, it is still possible to cite, for example, dynamic adaptation of network pressure to the power demand[2], as this has been frequently implemented on mobile or earth-moving machinery in a purely hydromechanical way (*load-sensing*). For control of power by throttling, the energy loss is proportional to the difference between the supply pressure and the pressure required by the load. One solution is to adapt, in real time, the supply pressure as required, by acting on the pressure setpoint of the pump compensator. In aerospace, this concept was evaluated in a system integration bench for the F-15 fighter [GRE 88, LAW 91] and in paper form regarding a European fighter [SPE 93]. It was eventually abandoned because it was difficult to implement in the context of aviation. To meet the dynamic requirements of the flight controls, it must in effect anticipate the power needs of actuators. This requires the installation of additional sensors and introduces complex couplings between computers, actuators and pumps.

3.4. Concepts combining hydraulics and electrics

The main idea here is to gain the most out of both technologies and the major advantages each technology has, by combining the two:

– from the hydraulics: the power density and ease of implementation for secondary functions to the actuators;

– from the electrics: the efficiency of control of power, the power density and the ease of re-configuring power networks.

3.4.1. *Local electro-hydraulic generation*

During a flight, a certain number of actuation functions are used infrequently, or may require a high level of power for only a short amount of time. In particular instances, these are some distance away from the main engines and their criticality often requires redundant supply by two separate energy sources. In conventional hydraulic power architectures, this situation requires the installation of highly specific or oversized hydraulic pipes, whose utilization rate is extremely low, and as such greatly penalizes mass. One solution is to locally produce, at a constant pressure, the hydraulic

---

2 Taking the example of an F-18 fighter [GRE 88], the flight records show that the supply pressure could be reduced by two-thirds during 93% of the time during a normal mission (only by 50% during a combat mission).

power from a pre-existing electrical power network, given that the electric power grids are available nearby as they are used for other systems (avionics, in-flight entertainment, etc.).

This solution, of which the principle is shown in Figure 3.2(b), has been implemented on the Airbus A380 [DEL 04] and helped remove one of the three main hydraulic systems otherwise found in conventional architectures. Figure 3.6 shows this new type of 2H–2E power architecture, i.e. it combines two hydraulic and two electric power networks. Three Local Electro-Hydraulic Generation Systems (LEHGS) provide a backup hydraulic power source to supply the steering system for the auxiliary landing gear and the braking system for the main landing gears. The backup generation is an integrated assembly composed of an electric pump, a reservoir, a hydraulic manifold, an accumulator and an electronic control unit. The electric pump consists of a brushless motor and a fixed displacement pump. The pump on–off function is controlled to hold the accumulator pressure in the operating range. The backup power supply is connected in parallel to the central power supply via a shuttle valve. It is engaged by the Braking and Steering/System Control Unit (BSCU) which controls a solenoid valve when it detects that the normal power channel has failed. The filling of the LEHGS tank is operated by the centralized hydraulic system. Therefore, the backup generation uses the same fluid as the centralized network to which it is connected.

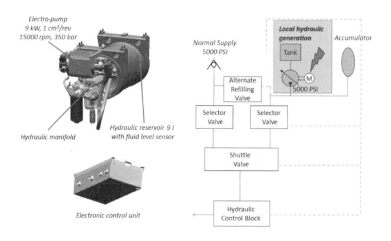

Figure 3.6. *Local electro-hydraulic generation [DEL 04] (see also Figure 7.8 in Volume 1). For a color version of this figure, see www.iste.co.uk/mare/aerospace2.zip*

3.4.2. *Electro-hydrostatic actuators*

This section briefly presents the different types of electro-hydrostatic actuators that will be covered in detail in Chapter 5. In the previous paragraph, the concept of local hydraulic generation simply produces an electrical power source solution for powering one (or more) conventional actuator(s). The control of power transmitted to the load to be actuated is still carried out by a servovalve.

By combining, for each actuator, a pump and cylinder, the concept of electro-hydrostatic actuator (EHA), shown in Figure 3.2(d) and 3.2(e), goes further:

– regarding *power metering efficiency*, it removes the power dissipation by throttling at the power stage of the servovalve. Functionally, power is controlled by sampling the electrical source of the only power required by the load. This is the principle of *power-on-demand*[3];

– it produces a *purely local hydraulic system*, even permitting for the removal of the centralized hydraulic power system[4]. This is a significant advantage, both for large commercial aircrafts, whose mass for hydraulic distribution is reduced, and for military aircrafts, whose survivability is improved;

– it retains the advantages of *hydraulics for power management functions*. The practical design of a cylinder inevitably introduces technological imperfections (friction at the seals and bearings, compliance of the fluid and walls, and mass of the movable portion). These imperfections have sufficiently little impact to allow for the transposition of mechanical power management functions to the hydraulic domain. Given that pressure is a good image of force and flow is a good image of speed, in many instances, required functions at the level of load (damping, snubbing, stabilization, protection against overload, clutching, braking, irreversibility) can be achieved through lightweight, compact and reliable hydraulic components (see Volume 1, Chapter 6).

---

3 As mentioned later on in the book, the technological imperfections and implementation of the secondary functions in practice generate additional power consumption at the source.
4 This is not entirely true in practice, see section 5.3.2.

3.4.2.1. *Principle of power on demand*

The cylinder (or possibly the hydraulic motor) is supplied with power by its own pump, itself driven by its electric motor. In the concepts adopted, an electronic Motor Power Drive (MPD) is inserted between the electric power source and the electric motor. Functionally, each of the power chain elements performs a power conversion function:

– the motor control electronics, if present, links the electric supply power variables[5] (current I_{DC} and bus voltage U_{DC}) to the electric output power variables (average current I and average voltage U).

$$\begin{cases} U = m\,U_{DC} \\ I = \dfrac{1}{m} I_{DC} \end{cases} \qquad [3.1]$$

where $m \in [-1; 1]$ is the modulation ratio;

– the electric motor links the electric power variables (current I and voltage U) to the mechanical rotation power variables (angular velocity ω and torque T):

$$\begin{cases} T = K_m\,I \\ \omega = \dfrac{1}{K_m} U \end{cases} \qquad [3.2]$$

The electromagnetic constant K_m [Nm/A or Vs/rad] is characteristic of the motor;

– the hydraulic pump, of the positive displacement type, links the mechanical rotation power variables with the hydraulic power variables (volume flow rate Q and differential pressure ΔP):

$$\begin{cases} \Delta P = \dfrac{1}{V_0} T \\ Q = V_0\,\omega \end{cases} \qquad [3.3]$$

where V_0 is the unit displacement of the pump (m³/rad);

---

5 These power variables are introduced in section 1.2.3 of Volume 1.

– the hydraulic cylinder links the hydraulic power variables to the mechanical translation power variables (output shaft speed v and force F developed on the load):

$$\begin{cases} F = S\,\Delta P \\ v = \dfrac{1}{S} Q \end{cases} \qquad [3.4]$$

where S is the effective hydrostatic area of the cylinder (m²).

Ultimately, the power is functionally transformed between the electrical and translational mechanical fields, by a factor SK_m/mV_0 or mV_0/SK_m according to the power variable being considered:

$$\begin{cases} F = \dfrac{SK_m}{mV_0} I_{DC} \\ v = \dfrac{mV_0}{SK_m} U_{DC} \end{cases} \qquad [3.5]$$

This result raises several important points:

1) *Power Conservation*: According to this purely functional vision, no technological imperfections are taken into consideration. The power drawn from the electric source is identical to the power output to the load, whatever the values of the transformation parameters, m, K_m, V_0 and S:

$$\mathcal{P} = UI = Fv \qquad [3.6]$$

Hence the name "*power-on-demand*".

2) *Hydrostatic loop*: The power transmission is performed partly through a loop consisting of the hydrostatic pump, the cylinder and the two hydraulic lines connecting them. Hence the name "*electro-hydrostatic actuator*". It will be seen in Chapter 5 that the power management functions carried out hydraulically somewhat increase the complexity of this hydrostatic loop whose principle is simple.

3) *Operation in four quadrants and regeneration*: If the chosen technology permits, the concept allows for regeneration: when the load is

aiding, it runs the motor in generator mode which feeds current to the power source.

4) *Fluid conditioning and power management*: In an electro-hydrostatic actuator, the implementation of a local hydraulic system does not exempt the incorporation of fluid conditioning functions (fluid reserve, filtration, refeeding) and power management (pressure relief, damped mode, etc.). These functions are typically integrated within a hydraulic manifold.

In terms of power metering, there is no technological solution to vary continuously and dynamically the power transformation ratio of the electric motor (K_m) and cylinder (S). This leaves only two options through acting by control on the force or speed that the actuator develops on the load to be moved:

– by acting on the pump displacement V_0;

– by acting on the modulation ratio m of the motor control electronics, if it exists.

3.4.2.2. *Action on pump displacement, EHA-VD*

The concept of a Variable Displacement Electro-Hydrostatic Actuator (EHA-VD) (see Figure 3.2(d)), or an IAP™ (Integrated Actuator Package) is based on an electric motor that is connected directly to the electric power source. The motor therefore always rotates in the same direction at an almost constant speed. This solution, shown in Figure 3.7, historically appeared when the motor control electronics were not yet sufficiently developed nor matured. It enables the power transmitted to the load to be controlled by acting on the pump displacement. Besides the absence of power electronics, this concept may even be applied without resort to any electronic transmission of information, thereby retaining a purely mechanical transmission of the pilot's commands to the motor displacement setting. Figure 3.8 shows a duplex EHA-VD aileron actuator, as developed by Lucas Varity (model C9106) and flight tested in the late 1980s on a Lockheed C-141 Starlifter military transport aircraft. Eventually, rapid advances in power electronics and the excessive heating of the fluid in the EHA-VD pump quickly led to abandonment of this concept, much to the benefit of the EHA-FD.

Power-by-Wire 77

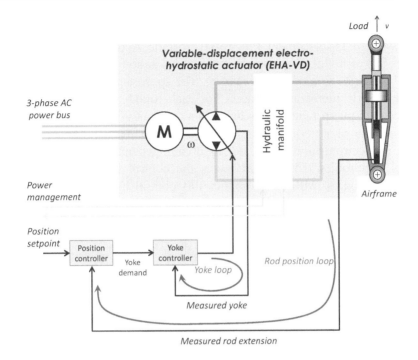

Figure 3.7. *Simplified architecture of an EHA-VD. For a color version of this figure, see www.iste.co.uk/mare/aerospace2.zip*

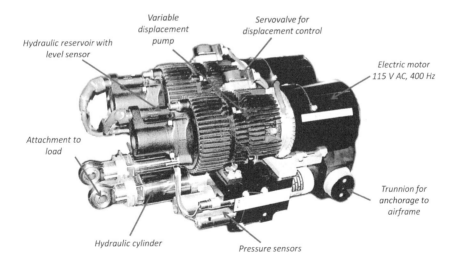

Figure 3.8. *Prototype of a duplex EHA-VD aileron actuator for Lockheed C-141*

3.4.2.3. *Action on the pump drive, EHA-FD*

In a Fixed Displacement Electro-Hydrostatic Actuator (EHA-FD), (see Figures 3.2(e) and 3.9), power is metered by action on the electric motor control through the modulation ratio m. The speed and direction of rotation for the electric motor are variable. For each round of the rotor, the fixed displacement pump functionally delivers a constant volume of fluid equal to $2\pi V_0$, on one of its ports A or B, depending on its direction of rotation. Power is electronically controlled, involving two main stages: the Motor Control Electronics (MCE) that issues the control signal and the Motor Power Electronics (MPE) that applies this control by acting as a variable transformer between the electric power source and the electric motor. Depending on the signal architecture of the aircraft, the control electronics of the EHA-FD is partially (motor control loop only, as shown in Figure 3.9) or totally (motor and position control loops) in charge of the servocontrol function of the load. Figure 3.10 shows the EHA-FD for the Airbus A400M.

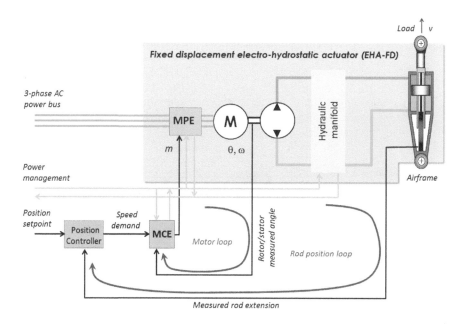

Figure 3.9. *Simplified architecture of an EHA-FD. For a color version of this figure, see www.iste.co.uk/mare/aerospace2.zip*

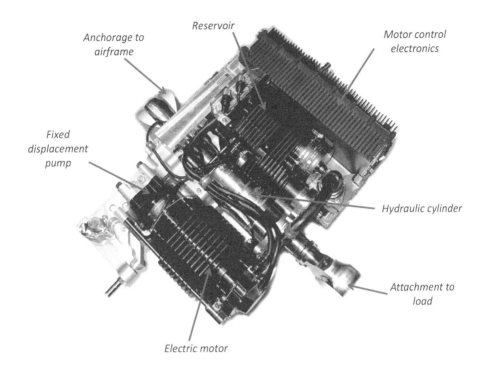

Figure 3.10. *EHA-FD actuator for the Airbus A400M*

3.4.2.4. *Hybrid metering, EBHA*

Within the same actuator, it may be interesting to combine the concept of ServoHydraulic Actuator (HSA), shown in Figure 3.2(a) (power metered by a servovalve and power supplied by a centralized hydraulic network) and EHA-FD concept, shown in Figure 3.2(e) (power controlled by action on the drive of a fixed displacement electric pump). This produces an actuator whose power supply (hydraulic or electric) and control (by servovalve or by action on the motor's power electronics) are both redundant and dissimilar (see Figure 3.11). This solution was introduced by Airbus on the A380. It is now used by other aircraft manufacturers like Gulfstream for its G650.

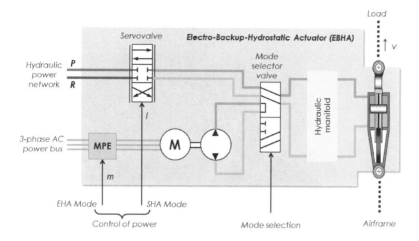

Figure 3.11. *Simplified power architecture of an EBHA. For a color version of this figure, see www.iste.co.uk/mare/aerospace2.zip*

In these assemblies, the EHA mode is used as a backup for the HSA mode: one mode being active at a time (as "exclusive or" functioning between the two modes). This is called an Electro-Backup-Hydrostatic Actuator (EBHA). The A380 implements eight EBHA: four for the rudder and four for the spoilers (see Figure 3.12).

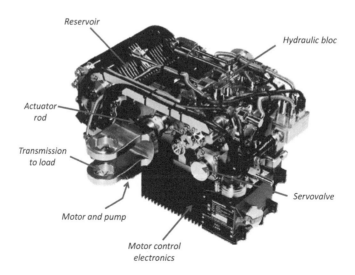

Figure 3.12. *EBHA spoiler for the Airbus A380, according to [BIE 04]*

Equally, it is potentially possible, by changing the design of the mode selector valve, to use both modes simultaneously, for instance to dispose of increased power ("and" function between the two modes). This is called the Electro-Assisted-Hydrostatic Actuator (EAHA).

3.5. All electric actuation (hydraulic-less)

3.5.1. *Principle of the electro-mechanical actuator*

The concept of the Electro-Mechanical Actuator (EMA), shown in Figure 3.2(f), replaces the hydrostatic transmission of the EHA with a mechanical transmission. The hydraulic fluid is abandoned as the power vector. Certainly in the field of aerospace, EMAs have long been used for low power. What is new, however, is their application in high power and critical functions. Chapter 6 will be dedicated to discussing this in greater detail.

Given the need for mechanical power at high force and at low speed, direct drive by an electric motor is not possible due to the constraints of weight and size. It is therefore essential to connect a mechanical gearbox to the electric motor. If the movement is linear, then a nut screw can produce a significant reduction and it may thus be possible to continue without additional gears. This is called a Direct Drive EMA. However, most EMAs integrate additional gears between the electric motor and the nut screw to help accommodate performance requirements and constraints. This is known as a Geared EMA. As the power management functions can no longer be transposed via hydraulics, they must be implemented in a mechanical, electro-magnetic or electric manner. The jamming of mechanical transmission is a feared event of which the consequences could be catastrophic if the actuator response to failure must be of the fail-passive type. This is a major reason why the EMA are still only rarely used for primary flight controls. They have nonetheless found their way into the secondary flight controls (actuation of slats, flaps, trim horizontal stabilizer), for primary flight controls of low quantity of service (space launchers) or for low power level (drones). Figure 3.12 shows the gear drive EMA used for the nozzle steering of the first stage of the Vega launcher (first flight in 2012).

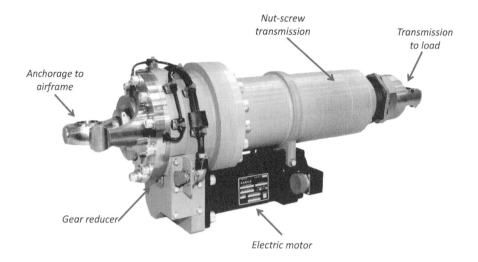

Figure 3.13. *Gear drive EMA for the P80 first-stage nozzle orientation of the VEGA launcher [DÉE 07]*

4

Electric Power Transmission and Control

The objective of this chapter is to provide a description of the needs, functions and architectures implemented in view of the development of the electric and electronic parts of PbW actuators[1]. This chapter will illustrate the constraints imposed on architectures and technology-driven functions according to the state of the art at a given moment. For this reason, this chapter follows a deductive approach, which leads to a presentation of the electric power architecture of PbW actuators starting from its needs and constraints, be they global or induced by the technologies employed. It is worth noting that the focus here is on providing, in a simple and synthetic manner, the elements that are relevant to system architects whose work interfaces that of experts in the design and command of electric machines. In order to obtain an expert perspective, the reader can refer to a wealth of specialist literature, such as [DED 11], [GIE 10], [GRE 97] and [RAS 11].

4.1. Electric power transportation to PbW actuators

Aircraft are fitted with several power sources that supply an electrical core from which power is distributed to users. The interest in this chapter is on PbW actuators supplied by a central power source. It is therefore logical to present below the means and constraints associated with the transportation of electric power to actuators.

---

[1] The author wishes to sincerely thank Marc Budinger, who offered valuable suggestions after reading this chapter.

4.1.1. *Form*

Similar to hydraulic power transmission under constant pressure and flow rate sharing, electrical networks carry power under imposed voltage and current sharing. The users are therefore connected parallel to the power network. The total current to be supplied to the network is the sum of currents drawn by the users. Power can be transmitted on the electrical networks and between the various elements of a PbW actuator in several ways:

a) Direct voltage (direct current or DC[2]). Functionally speaking, the bus voltage is constant. In the aviation industry, the standard values are 28 VDC for low powers and 270 VDC for high powers. The network requires two conductors, but the return line can be completed by the aircraft airframe itself, provided it is metallic, which leads to practically cutting the mass of the distribution network by half.

b) Alternating voltage (alternating current; AC). Functionally speaking, the bus voltage is sinusoidal, having a null mean value. In conventional solutions, the effective voltage is 115 VAC and the frequency is 400 Hz. The latest aircraft use variable frequency power generators, with frequencies ranging from 350 to 800 Hz (wild frequency), which have a higher efficiency. A three-phase network is commonly used, which consists of three sources of alternating voltage of the same frequency and with a phase difference of one-third of the period. If the network is balanced, meaning that the three voltages have the same amplitude relative to the neutral wire, the latter becomes redundant and it is possible to use only three conductors (the three phases) to transmit electrical power. This saves one conductor out of four.

Criteria such as mass, efficiency and technological maturity have favored electric power distribution in the balanced, three-phase alternating form for various reasons:

– the AC generators are simple and efficient;

– the AC voltage levels can be easily modified. This is done by a simple transformer that adapts the voltage level to the user's needs. In particular,

---

[2] It is worth noting that on certain power lines inside the actuator (those for direct current supply of the power electronics stage), the current linked to the force to be developed on the load can vary rapidly and reverse if the load becomes aiding. The DC acronym should therefore be interpreted as "direct voltage", rather than "direct current".

since the mass of the conductors in the network is determined by the current flowing through them, the high-voltage and low-current power transmission proves to be interesting;

– the electric arc produced in breakers when the current is interrupted is significantly weaker in alternating current than in direct current;

– in a three-phase alternating network, the power distribution is weaker (for a given power) and there are less conduction losses (for a given mass of conductors). For example, for an identical mass of conductors, the transmission of a given power through a three-phase network exhibits conduction losses that are four times smaller than that in a single-phase network.

Nevertheless, the AC solution presents some drawbacks compared with the DC one:

– the alternating current-carrying conductors act as electromagnetic interference emitters;

– the coupling of power sources is difficult, since they have to be synchronized in terms of frequency and phase;

– the component of reactive power generates currents on the power sources, which are not employed to effectively generate power, but are critical for the sizing of power generation and distribution.

4.1.2. *Voltage and current levels*

Table 4.1, which is associated with Figure 4.1, summarizes the conventions and basic relations of power transmission in the form of alternating current. In three-phase networks, there are phase voltages V (also called phase/neutral voltages or single voltages) and line voltages U (or phase/phase voltages, compound voltages). Similarly, there are line currents I and phase currents J. When the network is balanced, the three-phase voltages have the same amplitude and the sum of the phase currents I is null. The electric power that a conductor can convey is limited by the use restrictions relative to its limit values of current and voltage.

General relations for a sinusoidal power variable	
Instantaneous value	$x = x_0\sqrt{2}\sin(\omega t + \phi)$
RMS value, angular frequency, time, phase	x_0, ω, t, φ
Peak-to-peak amplitude	$x_{pp} = 2x_0\sqrt{2}$
Mean value	0
Frequency, period	$f = \omega/2\pi$ and $T = 2\pi/\omega$
Single-phase system	
Phase/neutral voltage	$V = V_0\sqrt{2}\sin(\omega t)$
Line current	$I = I_0\sqrt{2}\sin(\omega t + \phi)$
Powers: apparent, active, reactive	$\mathcal{P}_P = V_0 I_0$, $\mathcal{P}_Q = V_0 I_0 \cos\phi$, $\mathcal{P}_S = V_0 I_0 \sin\phi$
Balanced load three-phase system	
Star connection	$U_0 = V_0\sqrt{3}$
Delta connection	$I_0 = J_0\sqrt{3}$
Powers: apparent (S), active (P), reactive (Q)	$\mathcal{P}_S = \sqrt{3}U_0 I_0$, $\mathcal{P}_P = \sqrt{3}U_0 I_0 \cos\phi$, $\mathcal{P}_Q = \sqrt{3}U_0 I_0 \sin\phi$
Full-wave rectification AC→ DC	
Rectified DC voltage	$U_{DC} = \sqrt{2}V_0$ (single-phase), $U_{DC} = \dfrac{3\sqrt{2}\sqrt{3}}{\pi}V_0$ (three-phase)

Table 4.1. *Basic relations for power transmission in alternating form*

4.1.2.1. *Current*

Electrical conductors have an electrical resistance that is proportional to their length L and inversely proportional to their conductivity σ and cross-sectional area S. They generate a voltage drop ΔU that is proportional to their resistance R and to the current I that they carry:

$$R = L/\sigma S \qquad [4.1]$$

$$\Delta U = RI \qquad [4.2]$$

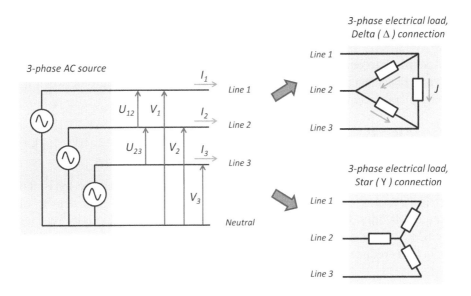

Figure 4.1. *Notations and conventions for three-phase electrical circuits*

On the other hand, their mass M depends on their density ρ, cross-sectional area S and length L:

$$M = \rho L S \qquad [4.3]$$

Similar to hydraulic power networks, when choosing electrical conductors (material and diameter), a compromise should be reached between the mass (minimizing the cross-sectional area) and the conduction losses (maximizing the cross-sectional area). The wiper is set by defining the maximum tolerable voltage drop for the conductor to fulfill its power transportation function. For example, a loss of 3.5% of the nominal voltage is acceptable for continuous operation, while twice this value is acceptable for intermittent operation (for a nominal voltage of 115 V, this means 4 and 8 V, respectively).

Power transmission conventionally uses copper wires: tin-coated for a maximum temperature of 150 °C, silver-plated for a maximum temperature of 200 °C or nickel-plated for a maximum temperature of 260°C. For high powers, the use of aluminum wires results in significant mass saving, as shown by the following calculation. Their maximum acceptable operating temperature generally ranges between 150 and 180°C.

According to equations [4.1] and [4.3], the product of linear mass density M/L and resistance per length R/L of a wire is equal to the ratio between its density ρ and its conductivity σ:

$$\frac{M}{L}\frac{R}{L} = \frac{\rho}{\sigma} \qquad [4.4]$$

This is an interesting product, as it reflects the impact that the mass and electrical resistance of conductors can have on the aircraft. The density of aluminum is only 30% of the density of copper, but its conductivity reaches only 63% of that of copper. For an equivalent energy loss, the mass of the conductor is reduced by 48% when copper is replaced by aluminum[3].

As mentioned previously, an important sizing criterion in electrical engineering involves the capacity to dissipate heat resulting from energy losses. Determining the size of the conductor to be used is therefore strongly influenced by its cooling capacity, which in turn depends on its environment: ambient temperature, altitude and other nearby conductors. These effects are taken into account by derating factors, as illustrated in Table 4.2 for the case of a 12 AWG (American Wire Gauge) tin-coated copper wire with a typical diameter of 2.8 mm. If a bundle is made of three conductors of this type for the three-phase supply with 115 VAC of equipment under the c) condition as defined in Table 4.2, the conductors reach a nominal temperature when an apparent power of 10.2 kW is supplied. Considering the conductor characteristics (linear mass density of 30.5 g/m and resistance per length of 6.7 Ω/km at 150 °C), the bundle has, under these conditions, a linear mass density-to-power ratio of 8.9 g/m/kW. It dissipates a power of 17.7 W per unit length or 0.17% of the carried power and it generates a voltage drop of 4 V for 20 m of wire. These values can be compared with those provided by Table 6.2 in Volume 1 for rigid hydraulic pipes. The same power is conveyed by a dash-8 titanium pipe through which a type IV fluid flows at 207 bar with a speed of 5 m/s. The linear mass density-to-power ratio is already 20.6 g/m/kW without counting the return piping. The loss amounts to 21 W/m at 38 °C, but

---

3 Other drawbacks should also be considered: the coefficient of linear thermal expansion increases by 35% and mechanical strength is reduced by 36% when copper is replaced by aluminum.

it reaches approximately 1 kW/m at −54 °C! These orders of magnitude clearly show the advantages of power transmission in electrical form.

It is worth noting that the most unfavorable case for the definition of an electrical wire arises when the wire temperature is high, since copper conductivity decreases[4] (it is reduced by half when the temperature rises from −54 °C to 178 °C). By contrast, the sizing case in hydraulics emerges at the lowest temperatures, since the fluid viscosity increases rapidly (for a type IV fluid, it typically increases twofold when the temperature decreases by 15°C between 50 and −50°C).

12 AWG tin-coated copper wire, rated temperature 150°C			
Successive effects considered	Conditions	Current for rated temperature (A)	Derating factor
Reference	Single conductor Ambient temperature 25°C Zero altitude	56 A	1
a) Ambient temperature	Single conductor Ambient temperature 70°C Zero altitude	46 A	x 0.821 → 0.821
b) Wire bundling	3-wire bundle Loaded at 100% Ambient temperature 70°C Zero altitude	34.5 A	x 0.75 → 0.616
c) Altitude	3-wire bundle Loaded at 100% Ambient temperature 70°C Altitude of 10,000 m	29.6 A	x 0.86 → 0.53

Table 4.2. *Example of electrical wire characteristics*

4.1.2.2. Voltage

In a similar manner to the operating pressure in hydraulics, the level of acceptable voltage is imposed at a given moment by the state of the art: for PbW, these are the state of the power electronic components and the resistance of insulators to partial discharges.

Three-phase electric power transmission through alternating voltage with an effective value of 115 VAC is a standard that has been well established

---

4 But the factor of exchange with the environment increases, which is favorable.

for several decades. For high-power applications in direct current, the standard value is 270 VDC (or +/− 135 V), which is a natural result of the full-wave rectification of three-phase 115 VAC (according to Table 4.1). Unfortunately, the increase in electric power to be transmitted is no longer fitted to such levels of voltage, since the resulting increase in currents directly impacts the mass of conductors. This is why some recent aircraft models such as the Airbus A350 have started to implement three-phase electrical networks at 230 VAC, which is twice the conventional voltage, hence the term *double voltage*. The levels of direct voltage can also be increased in order to form HVDC (High-Voltage Direct Current) networks. For example, the full-wave rectification of a three-phase voltage of 230 V yields a direct voltage of 540 (or +/− 270) VDC.

At the level of power electronics, the increase in supply voltage is made possible by the advances in the field of wide band gap semiconductors based on silicon carbide (SiC). Compared with common silicon-based components, these components typically withstand a double voltage (1,200 V vs. 600 V). Their switching time and their switching and conduction losses are significantly reduced, while their operating temperature range is expanded. Unfortunately, when used in commutation for power control by pulse-width modulation, these components yield steep-front voltages in the electrical wires (typically between 3 and 5 kV/μs), followed by oscillations that yield overvoltage [MIH 11]. When the potential increases beyond a certain threshold (Partial Discharge Inception Voltage or PDIV), of the order of 600 V, the resulting magnetic fields ionize the ambient air. If the conductors are too close to one another, ionization finally leads to partial discharges that are perceived as bluish light, which is the Corona effect illustrated in Figure 4.2. Partial discharges appear more easily when the pressure and the distance between conductors decrease (this relation is described by Paschen's law). Being located outside the pressurized area, motors and local electronics of actuators are particularly exposed to the risk of partial discharges at high altitude. Given that their winding turns are separated only by their insulator, motors can experience Corona effects if their impregnation is flawed due to residual air cavities. The latter are harmful to reliability because they considerably accelerate the ageing of insulators [KOL 08].

Figure 4.2. *Corona effect in an electric motor winding [COU 16]. For a color version of this figure, see www.iste.co.uk/mare/aerospace2.zip*

4.2. Electric motors

4.2.1. *Elementary electric machines*

This section briefly reviews the operation of elementary electric machines in order to facilitate a better understanding of their imperfections and requirements in terms of associated electronics.

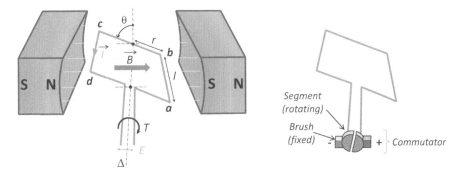

Figure 4.3. *Elementary electric machine (left) and the principle of current commutator (right). For a color version of this figure, see www.iste.co.uk/mare/aerospace2.zip*

In Figure 4.3, the diagram on the left presents a simplified and generic way to convert mechanical power into electrical power and vice versa in a rotating electric machine. A rectangular coil of length l and width $2r$ carries a current I. The coil rotates around the axis Δ, its rotation being defined by an angular parameter θ relative to the machine frame. By means of elements inside the machine, such as permanent magnets, a magnetic field B is generated between the pair of poles N (North) and S (South).

The combination of current I flowing through the coil and magnetic field B to which the coil is subjected generates an electromagnetic force F (or Lorentz force) that is perpendicular to the direction of the current flow and to the magnetic field:

$$\vec{F} = I \int_{coil} \vec{dl} \wedge \vec{B} \qquad [4.5]$$

Under these conditions, only the electromagnetic forces F_{ab} and F_{cd} developed on the segments ab and cd of the coil generate an electromagnetic torque T along the axis of rotation Δ:

$$T = r \sin\theta (F_{ab} + F_{cd}) \qquad [4.6]$$

where $F_{ab} = F_{cd} = F = I\,l\,B$

This finally leads to

$$T = BS \sin\theta\, I \qquad [4.7]$$

where $S = 2rl$ is the area of the loop.

Furthermore, the motion of the coil in the magnetic field B induces a voltage E across the coil terminals under the effect of the rate of variation in the magnetic flux $d\phi/dt$ through the coil. This effect is modeled by Faraday's law:

$$E = -\frac{d\phi}{dt} \qquad [4.8]$$

where $\phi = BS \cos\theta$ is the magnetic flux across the coil.

The previous equations finally lead to the formulation of the two equations describing the electromagnetic transformation of power between the electrical domain (power variables E and I) and the mechanical domain (power variables T and ω). It then leads to:

$$\begin{cases} T = K_{ei} I \\ \omega = \dfrac{1}{K_{ei}} E \end{cases} \qquad [4.9]$$

where $\omega = d\theta / dt$ is the angular velocity of the coil relative to the frame of the machine and $K_{ei} = BS\sin\theta$ is the instantaneous power transformation ratio (the mean value K_m is called electromagnetic constant or torque constant).

NOTE.– Lorentz's and Faraday's laws are well suited to analytical calculations. Another approach is nevertheless interesting, because it facilitates the qualitative understanding of the operation of electric machines. For this purpose, the electric wire is considered an electromagnetic coil whose axis is perpendicular to the coil plane. Owing to the current flowing through the coil, a magnetic field is generated along this axis. The electromagnetic torque results from the interaction of the magnetic fields B_s and B_r created at the stator level (by the permanent magnets in Figure 4.3) and at the rotor level (by the coil in Figure 4.3), respectively. It reaches a maximum value when the two fields are perpendicular to one another, and it is null when the two fields are parallel:

$$\vec{T} = k \vec{B}_s \wedge \vec{B}_r \qquad [4.10]$$

In the case of electric machines, the focus is generally on increasing the electromagnetic constant, which means generating a higher torque for a given current, for a motor, and generating a higher voltage for a given velocity, for a generator. This can be achieved by increasing the number of turns N, the number of poles p, the area S of the turn (by increasing the active length l or the radius of action r) and the field B (using ferromagnetic materials). In the elementary example provided in Figure 4.3, the mechanical angle θ (rotor/stator relative angular position) is equal to the electrical angle δ that defines the orientation of the turn relative to the magnetic field. For a stator with p pairs of alternated poles, the electrical angle δ is

$$\delta = p\theta \qquad [4.11]$$

and the instantaneous electromagnetic constant can be written as

$$K_{ei} = NBS \sin p\theta \qquad [4.12]$$

In the elementary machine shown in Figure 4.3, the instantaneous transformation ratio is a sinusoidal function of the electrical angle. In order to generate a constant direction torque (or to generate a positive voltage for a given rotation direction), the current should therefore reverse when the electrical angle is a k integer multiple of 180°:

$I = I_0$ and δ ranges between $(2k)\pi$ and $(2k + 1)\pi$

$I = -I_0$ and δ ranges between $(2k + 1)\pi$ and $(2k + 2)\pi$ [4.13]

For example, in a synchronous electric motor with two pairs of poles, the electrical angle δ varies by 360° when the rotor turns by a mechanical angle of 180°. If the motor turns at 12,000 rpm, the frequency of the current to be produced in order to supply the motor is 400 Hz.

4.2.1.1. Direct current machines (or DC motor)

The current is transmitted to the rotating coil by the stationary brushes and it is inverted by means of a segmented commutator placed on the rotor[5], as shown on the right of Figure 4.3. To take better advantage of the magnetic field, the rotor can be fitted with several windings separated by an angular distance. The segmented commutator then serves to supply the windings whose electrical angle is best fitted to torque generation, which means the one for which sin(δ) is maximized.

4.2.1.2. Alternating current synchronous machines

An alternating current is carried into the coil by stationary brushes, by means of two slip rings on the rotor. The applied current has to reverse in synchronicity with the electrical angle of the machine. It is possible to do without brushes and collector rings, by reversing stator and rotor: the turns are stationary (on the stator) and the magnetic field is generated by magnets (located on the rotor). This is called a Permanent Magnet Synchronous

---

5 As indicated in Volume 1, section 4.3.4, an analogy can be drawn here with the valve plate of a pump or of a hydraulic motor with axial pistons, which performs the function of active chamber commutation in correlation with the rotor angle.

Machine (PMSM). The task of current commutation in correlation with the electrical angle of the machine is fulfilled by power electronics. These are generally three-phase machines and have windings with star connections (phase currents are equal to line currents) (see Table 4.1).

4.2.1.3. *Alternating current asynchronous machines*

The operating principle of asynchronous machines differs to some extent from the basic principles described previously. The stator or inductor consists of inductive coils supplied by alternating current. These coils generate an electromagnetic field (inductor field) that rotates relative to the motor case. The rotational frequency is equal to the electrical frequency of the current across the coils divided by the number of pairs of poles of the inductor (for example, 1,500 rpm for an electrical frequency of 50 Hz on a machine with two pairs of poles). The rotor consists of elementary turns (squirrel-cage) in which the current is induced by the difference in angular velocity between the rotor and the inductor magnetic field. This current generates a magnetic field due to which the rotor behaves as a magnet in relation to the inductor field. The electromagnetic force thus generated drives the rotor, which closely follows the inductor rotating field: the driving torque depends on the relative slip s, or in other terms, on the relative angular velocity ($\omega_e - \omega_s$) between the rotating field and the rotor:

$$s = \frac{\omega_e - \omega_s}{\omega_e} \qquad [4.14]$$

Close to synchronism speed, this torque is proportional to the slip.

4.2.2. *Conversion of electrical power into mechanical power*

Contrary to hydraulic actuators, when presenting PbW, it is difficult to dissociate the power metering functions (power electronics) from those of electromechanical conversion (electric motors), since they are intertwined. Indeed, the strategy of power metering executed by power electronics relies on motor design and on its technological implementation. As mentioned at the beginning of the chapter, the objective here does not include a detailed account of the design, modeling and command of electric machines. There are numerous works that provide in-depth presentation of these subjects. The focus here is on generic architectural and operating aspects, as well as on technological imperfections and constraints related to the use of electric power in aerospace actuation. Therefore, the starting point relates to the requirements at the level of the driven load. This offers the possibility to define the best-suited types of electric motor and, as a next step, to further elaborate on the requirements they entail in terms of power electronics.

There are numerous electric motor concepts and their classification relies on different unifying criteria. Figure 4.4 presents an example of synthetic classification.

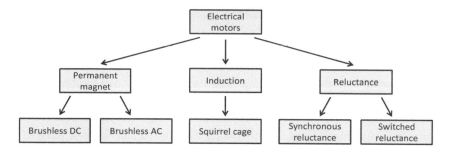

Figure 4.4. *Various electric motor concepts, according to [CAO 12]*

The motors used in aerospace power actuation have to meet highly specific requirements and constraints (according to Chapter 1 of Volume 1):

– Closed-loop operation. A vast majority of applications require load-position control: mobile surfaces for flight controls and orientation of the wheels of the auxiliary landing gear for ground steering. The force control refers essentially to the braking of the main landing gears. There are few applications referring to speed control: the *green taxiing* system can be cited as an example, currently being the object of several development projects. Moreover, the closed-loop control has become a requirement for functions that were until now point-to-point hydraulically controlled (deployment of thrust reversers, extension/retraction of landing gears): on the one hand, for secondary functions such as speed limitation and deceleration when approaching end-stops, and on the other hand, for motion control and minimization of the transient power drawn from the source.

– High power density and efficiency. Motor mass and efficiency have a snowball effect on the whole upstream power system, up to the aircraft itself (wing surface area for the aerodynamic lift, engines for the drag generated by the wings and for secondary power supply, etc.). Equation [4.7] has shown that the torque yielded by an electric motor is a direct function of the current across the windings, the number and the active length of the turns and the magnetic field to which they are subjected. The increase in current leads to an increase in the mass of conductors, which are chosen based on diameter and material criteria in order to minimize heating, power losses by conduction and overall external dimensions. The increase in magnetic field is equivalent to the increase in the mass of magnets and laminations. For a given power requirement, it is therefore interesting, in terms of mass, to generate low torque at high speed. Speed is in turn limited by various effects (stress due to centrifugal forces, core losses, friction, etc.). Beyond motor efficiency at rated power, the mean efficiency on a generic mission should be considered, which reflects the heat dissipated through energy losses: effective losses are a major driver for the sizing of the whole electric chain.

– Low inertia. Since most applications are dynamic, the closed-loop bandwidth and the energy used[6] by a motor during a mission are closely related to the inertia of its mobile parts. Moreover, rotor inertia often predominates over other types of inertia (mechanical transmission and load). Therefore, it has a direct influence on stability (for example, flight controls *flutter* or steerable landing gear *shimmy*), *force-fighting* or transient forces during arrival to mechanical stop.

– Service life and environment. The service life of a motor shall span over several tens of thousands hours (or cycles) and its operating environment is extremely harsh (humidity or water splashing, salt spray, explosive atmosphere, etc.).

The above considerations lead quasi-exclusively to using brushless machines. From this perspective, asynchronous machines should seem well adapted. Unfortunately, their performances at low speed and their torque density are poor, which leads to their dismissal from aerospace actuator applications. Even though variable reluctance machines are still under development, permanent-magnet machines are at the moment best adapted for actuation needs. The windings are located on the stator, which facilitates the outward dissipation of the heat yielded by various energy losses inside the motor. The use of high-performance magnets (for example, samarium–cobalt magnets) generates a significant increase in the torque density due to

---

6 In the absence of regeneration, which is generally the case.

the strong magnetic field they generate. To get an idea of the order of magnitude, Parvex industrial motors can serve as illustration: permanent torque density typically amounts to 0.25 Nm/kg for a DC motor with ferrite magnets, it doubles when samarium–cobalt magnets are used, and it doubles once again when a brushless motor is used instead of a brush motor. When electronic commutation replaces the mechanical commutation of the segmented commutator, there are many ways to optimize power control with high efficiency and to expand the field of operation, for example, by flux weakening. The class of permanent magnet brushless motors comprises in practice:

– BLDC (BrushLess Direct Current) motors, which operate similarly to brush direct current motors but employ electrical commutation instead of the segmented commutator;

– PMSM motors, which operate similarly to an alternating current synchronous motor, but without brushes.

4.3. Power conversion, control and management

4.3.1. *Electric power system of a PbW actuator*

The electric power system of a PBW actuator implements various power conversion functions, either fixed or variable. On the one hand, these functions have to adapt the form (alternating/direct), the voltage or the frequency of the power source to the needs of the motor. On the other hand, they must allow for high efficiency power control, from a command signal that allows the optimum exploitation of the motor performances. The converters[7], as these elements are usually called, are now made in a static form, from power electronics components (diodes, transistors, thyristors). Figure 4.5 illustrates a classification of these elements, which is based on the nature of the conversions they perform between the alternating and direct voltage domains:

– rectifiers convert a single-phase or multiphase alternating form into a direct form;

---

7 In order to distinguish the modern converters with power electronic components from the converters using rotating machines, the term "static converters" is employed.

– inverters convert a direct form into a single-phase or multiphase alternating form;

– dimmers modify the RMS voltage of an alternating source without changing its frequency. When using matrix converters, both frequency and phase can be modified, eliminating the need to use a DC-link for brushless electric motor control [WHE 07];

– choppers, associated with a low-pass filtering most often generated by the load itself, modify the voltage level of a direct voltage source.

The thick arrows in the figure symbolize the functional paths of power, the thin arrows symbolize converter command signals, while the symbols represent the function without an indication of how it is executed. Referring to Figure 4.5, an interesting aspect that is worth noting is the *DC-link* which is the commonly used power architecture for the control of a brushless motor from an AC power source. The conversion to the DC domain is performed by a rectifier, followed by a new conversion to the AC domain, by means of an inverter. Passing directly through a matrix converter [EMP 13] has not yet reached the level of technological maturity that would allow it to be used in aerospace actuation.

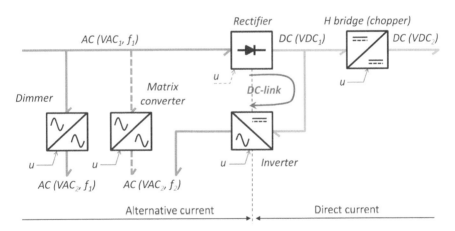

Figure 4.5. *Various power electronic converters. For a color version of this figure, see www.iste.co.uk/mare/aerospace2.zip*

There are several aspects that are worth considering as they have a strong impact on the electronic architecture:

a) The nature of the actuated load and of the power source. For an element of power transmission, four power quadrants can be defined, according to the sign of the variables of its output power[8] (for example, for an actuator: torque transmitted to the load and load speed, or still for a chopper: current and voltage supplied to the motor). When the product of power variables is positive, the element operates in a motor quadrant: there is an *opposite load* [9] and power is transmitted from the element to the load. Conversely, when the product is negative, the element operates in a braking quadrant: there is a *driving* or *aiding load* and power is transmitted from the load to the element under consideration. If during the mission the load can be aiding, the power system should be able to accept receiving power from the load, and store, dissipate or inject it back to the power source, as applicable. In Figure 4.5, the arrows indicate the functional paths of power. However, provided that they are designed for this purpose, the majority of converters can operate in the four power quadrants.

b) Disturbances induced in the power variables by the coupling of various elements and by the parasitic phenomena that they generate because of their imperfections. In order to limit the effect of these disturbances, filters should be inserted between the elements [FOC 11]. It is obvious that these disturbances should first of all be limited by the design that is appropriate for each element.

c) The necessity to control the converters. There are few converters that can execute the conversion function in the absence of control. This is, for example, the case of diode rectifiers. Nevertheless, most converters require control electronics.

4.3.2. *Principle and interest of static switches*

In hydraulic actuation, power control typically involves acting upon a variable hydraulic resistance placed on the power path: the supply pressure that exceeds the needs of the load is eliminated as head loss across the terminals of this resistance. The power drawn is hence identical to the power

---

8 See Table 1.3 in Volume 1.
9 The term resistive load is also used, but it should be avoided, as it may mislead the reader into thinking that the load is purely dissipative from a power point of view.

required to develop the maximum force; therefore, the energy efficiency of the control principle is low. In power electronics, the interest lies in the possibility to develop high-speed static switches (switching time below 1 μs), which exhibit a very low resistance R_{on} in forward conducting mode (some 10 mΩ), a very high resistance R_{off} in forward blocking mode (>1 MΩ) and low energy required for commutation (several mJ per change of state) at the same time. Power is then controlled by Pulse-Width Modulation (PWM), which means rapidly switching (several kHz to some 10 kHz) between conducting/blocking states by varying the mean duration t_{on} of the conducting state relative to the mean duration t_{off} of the blocking state. The mean voltage applied to the load is then controlled by acting on the duty cycle m, which varies between 0 and 1:

$$m = \frac{t_{on}}{t_{on} + t_{off}} \qquad [4.15]$$

The load connected to the static switch behaves in general as a low-pass filter. This is particularly the case of motors with windings that have inductive and resistive effects: although the voltage applied by on/off commutation appears as a sequence of steps, the current across the load is less pulsed thanks to this low-pass effect. In practice, there are several principles for commutation control in PWM [RAS 11], each having different advantages in terms of complexity and current ripple magnitude. The best known consists of activating the on-state at regular intervals by means of a constant frequency clock. It is the triangular carrier PWM and its principle is presented in Figure 4.6.

Figure 4.7(a) presents the ideal characteristic of a switch (R_{on} = 0 and R_{off} = ∞). Figure 4.7(b) shows the ideal characteristics of a power electronics-based static switch. Given the power levels of aerospace actuators, in most cases, this switch consists of an IGBT (or possibly a MOSFET for lower currents) that works in commutation mode. The commutation command is activated by the increase in the gate current (or in the gate voltage for a MOSFET). It is a fact that transistors do not allow reverse voltage or current across their terminals: they can only operate in the first power quadrant. Therefore, in order to authorize reverse currents during transient commutation or under aiding load conditions, they are connected to an antiparallel freewheeling diode. Figure 4.7(c) illustrates the plot of the real characteristic of a static switch. The slopes of the real characteristics

$I_s(U_s)$ corresponding to conductances $1/R_{on}$ and $1/R_{off}$, as well as the threshold voltages of the diode (U_d) and of the transistor (U_t) in forward conducting mode, are worth noting, as well as the fact that the switch obtained is never off when reverse voltages are applied.

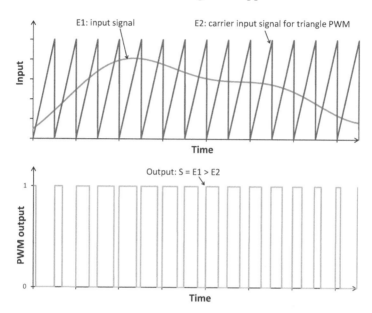

Figure 4.6. *Principle of the triangular carrier PWM*

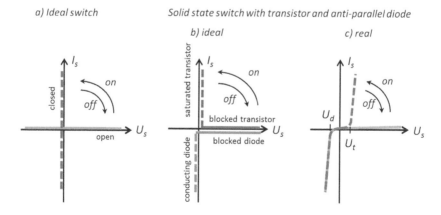

Figure 4.7. *Characteristics of a static switch. For a color version of this figure, see www.iste.co.uk/mare/aerospace2.zip*

4.3.3. *Groups of switches: commutation cell, chopper and inverter*

As shown in Figure 4.8, static switches can be grouped depending on the needs of the function to be developed. The IGBT are then integrated and embedded in modules, as shown in Figure 4.9.

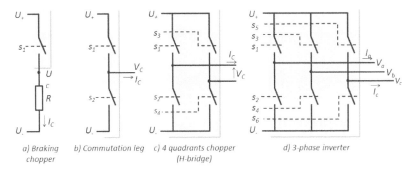

Figure 4.8. *Groups of static switches*

Figure 4.9. *IGBT symbol and example of IGBT modules in the power drive electronics of a large civil aircraft EBHA. For a color version of this figure, see www.iste.co.uk/mare/aerospace2.zip*

a) Dynamic brake chopper:

It will be seen that it may be required to dissipate power in a purely resistive load. For this purpose, a dynamic brake chopper is developed, which consists of connecting to the DC voltage source a serial association of static switch and a power resistance of value R. The mean power $\widehat{P}_t$ dissipated in the resistive load is controlled, varying the duty cycle of the static switch command:

$$\widehat{P}_t = m \frac{(U_+ - U_-)^2}{R} \qquad [4.16]$$

b) Commutation cell:

The association of two switches in series gives rise to a *commutation cell (leg)*. This leg serves to apply a voltage U_C across the load equal to the positive supply voltage U_+ or to the negative supply voltage U_- according to the states s_1 and s_2 commanded for the upper or lower static switch. Control logic does not allow the switches to go on simultaneously, in order to avoid the short circuit of the voltage source. The load is therefore under mean voltage:

$$\hat{V} = m(U_+ - U_-) \tag{4.17}$$

where m ranges between 0 and 1.

The commutation cell can control power in two quadrants (positive or negative load current I), if this is allowed by the power source.

c) Four-quadrants chopper:

To authorize the load (for example, a brush DC motor) to operate in all four power quadrants, two commutation cells are grouped in a *chopper*. Each cell supplies one of the power wires of the load. Adding a second commutation cell allows for reversing of the load voltage, which is expressed by the relation 4.17, where m ranges between -1 and 1.

d) Inverter

In the case of an AC load with n-phases (for example, a three-phase BLDC motor), an *inverter* is obtained by grouping n commutation cells that elaborate the n-phase voltages (V_a, V_b, V_c)[10] from the direct voltage source. In this case, the mean value of each phase voltage can vary between U_+ and U_-.

Finally, it is important to keep in mind that power is controlled by modifying the voltage imposed across the load by high frequency on/off commutation, controlled by logic signals applied to the switches. The current across the load depends on the response of the load to the voltage applied

---

10 For the control of three-phase motors, the labels a, b and c are usually associated with the supply lines of the motor. However, from a power perspective, these lines are often identified by the labels U, V and W.

and it is smoothed by its low-pass effect. Finally, these converters comprising transistors and power diodes are imperatively supplied by direct voltage sources, which explain the *DC-link* presence in Figure 4.5.

4.3.4. *Inverter command*

Motor power electronics (MPE) receive the commutation commands elaborated by motor control electronics (MCE). The functions of elaboration and transmission of these commands have to satisfy several constraints:

– Isolation. The control part operates at low voltage. It has to be isolated from the power part for switches control since it contains sensitive components (processors, etc.). The isolation can, for example, be achieved by transformers with a high-frequency carrier or by optocouplers.

– Conditioning and amplification. Though the energy for transistor command is low, the power part integrates functions for the conditioning and amplification of signals collected after command coupling in order to adapt them to the switch control needs.

– Exclusivity. As mentioned previously, it should be impossible to simultaneously close both switches on a leg, as this would short-circuit the DC power supply. Moreover, static switches exhibit delayed closing or opening, which requires the introduction of a delay between the commands of change of state of the two switches on the same leg.

The on/off control of the switches of a converter associated with a motor is particularly interesting, as it offers the designer many possibilities (respecting the exclusivity condition). On the one hand, it has to generate an inductor field with a precise orientation relative to the field generated by the rotor magnets: it is the commutation function. On the other hand, it has to adjust the module of the rotating field in order to optimally control the electromagnetic torque generated by the motor.

a) Static (or six-step) control of a BLDC motor:

The static control of a permanent magnet motor is an electric version of the mechanical commutation of brush DC motors. Its principle is presented in Figure 4.10. Static control is easy to implement, as it relies on a static model (or steady regime) of the motor. On the one hand, it only uses discrete data on the rotor/stator angle θ. The stator shall simply be fitted with sensors, for example, Hall effect sensors, which detect the presence of the magnetic field generated by the rotor magnets. On the other hand, the six-step control requires no calculation since the logic commands of commutation of switches are a combinatorial function of logic signals provided by these sensors. It is worth noting in Figure 4.10 the three-phase, balanced alternating form rebuilt by the control, depending on the motor angle for phase voltages V_a, V_b and V_c (as the motor has two pairs of poles, an electric cycle corresponds to one half rotor revolution relative to the stator). To minimize the torque ripples, the motor should be designed in such a way that its ElectroMotive Force (EMF) is a trapezoidal function of the electrical angle (which differentiates the BLDC motor from the PMSM motor). This characteristic is generally obtained with magnets placed on the rotor surface generating non-salient poles.

Unfortunately, this type of command does not take full advantage of the degrees of freedom available for controlling the various switches of the inverter. Motor torque ripple increases at high speed, the inductor field generated by the current induced in response to the voltage imposed by the inverter is not optimally used and the speed is limited by the DC bus voltage.

b) Vector control (field-oriented control or FOC):

When used with an adapted motor, the vector control technique limits the previously mentioned downsides by generating sinusoidal phase voltages that have no fronts. Performances are far better, but complexity increases. The motor type must be PMSM, which means that it has to be designed to generate a sinusoidal electromotive force as a function of the electrical angle. This is generally done with salient poles (for example, using buried magnets). Vector control also requires precise position measurement (several tens of points on an electrical cycle) and inverter control requires stronger computation and real-time signal processing capacity.

Electric Power Transmission and Control 107

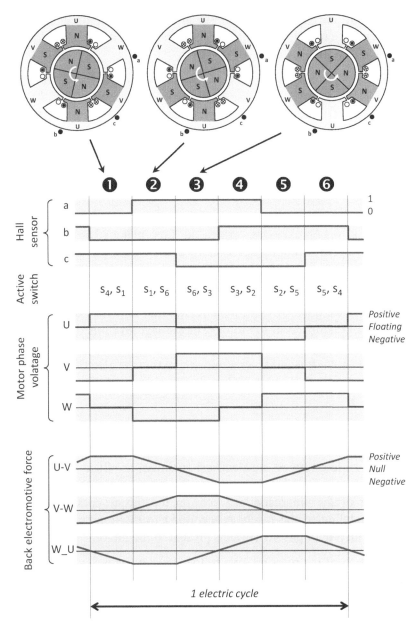

Figure 4.10. *Example of six-step control of three-phase BLDC motor (concentrated winding, six teeth and two pairs of poles) by a three-phase inverter based on three Hall effect sensors spaced at 120° electrical degrees, adapted from [JIA 14]*

In view of the modeling, analysis and elaboration of a three-phase machine vector control, several reference frames and transformations are defined, as illustrated in Figure 4.11:

– From a physical point of view, the inverter generates three-phase voltages (V_a, V_b, V_c) that give rise to three-phase currents (I_a, I_b, I_c). These components are represented in the plane perpendicular to the rotor/stator axis of rotation and whose origin O is located on this axis, along the axes (a, b, c) distanced at 120°.

– From a space point of view, a plane frame of reference of directions (α, β) and origin O is linked to the stator. The voltage vector $\vec{V}$ and the current vector $\vec{I}$ resulting from the vector addition of their respective components along the (a, b, c) axes are drawn in this plane. These vectors are characterized by their amplitudes V and I and by their phases. The passage from the three-phase basis (a, b, c) to the two-phase orthogonal basis (α, β) constitutes the *Clarke transform*, illustrated in Figure 4.11(a).

– From a space point of view, a plane frame of reference of directions (d, q) and origin O is linked to the rotor. The passage from the (α, β) frame of reference to the (d, q) frame of reference is achieved by a counterclockwise rotation by θ angle. It should be reminded that θ is the electrical angle of the stator relative to the rotor. This is *Park transform*, illustrated in Figure 4.11(a).

These two transforms explicitly evidence the torque control angle ψ and the phase difference angle φ between voltage and current. The use of the (d, q) basis is interesting, as the electrical quantities are fixed in this basis for a given operating point (torque/speed). This greatly facilitates the development of the command that acts on the torque control angle. The electrical modeling of the machine according to this representation allows for the introduction of physical effects present in the motor:

– the motion of the rotor relative to the stator generates an electromotive voltage $E = K_m \omega$ ($\omega = d\theta/dt$) that is carried by the axis q (since it is the axis with the highest flux change);

– the resistive effect of the wires in windings, whose resistance is R, generates a voltage drop proportional to the current;

– the inductive effect of windings, whose inductance is L, generates a voltage drop proportional to the rate of current change.

a) Clarke and Park transforms

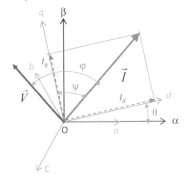

b) Phasor diagram (general case)

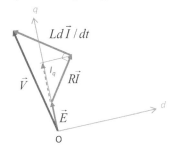

c) Field oriented control: min copper losses (max torque for a given current)

d) Field oriented control: flux weakening (max speed at given supply voltage)

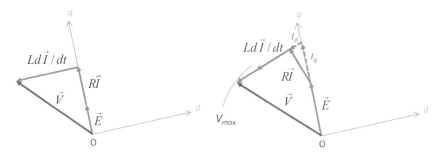

Figure 4.11. *Principle of vector control for PMSM. For a color version of this figure, see www.iste.co.uk/mare/aerospace2.zip*

Overall, the resultant voltage applied to the motor can be written as:

$$\vec{V} = K_m \omega \vec{q} + R\vec{I} + L\frac{d\vec{I}}{dt} \qquad [4.18]$$

This vector sum can be graphically represented by a *phasor diagram*, as illustrated in Figure 4.11(b). Since $\vec{I}$ is a rotating vector, the vector $d\vec{I}/dt$ is perpendicular to it. In the frame of reference (d, q), the components of the current vector $\vec{I}$ are I_d and I_q. Its decomposition along the (a, b, c) axes yields the phase currents (I_a, I_b, I_c).

The vector control principle relies on the possibility of separately controlling two quantities, as shown in Figure 4.11(b):

– the direct component I_d of the motor current that does not generate electromagnetic torque, as it produces an inductor magnetic field parallel to the field produced by the rotor magnets;

– the quadrature component I_q that generates the electromagnetic torque, as it produces an inductor magnetic field perpendicular to the field produced by the rotor magnets.

In other terms, it can also be said that the vector control serves to control the motor current I and the phase difference between the motor current and voltage. Owing to this degree of freedom, motor performances can be enhanced by producing the required effect in an optimal manner:

– In most cases, the focus is on generating maximum electromagnetic torque for a given current. In this case, the voltage provided by the inverter to the motor should allow for maintaining $I_d = 0$ (or the equivalent $\psi = 0$) in order to minimize the windings' copper losses, as shown in Figure 4.11(c).

– In contrast to DC or BLDC motors, the maximum speed is not limited by the supply voltage. In effect, it is possible to vary the direct and quadrature currents in order to raise the speed limit: this is called *flux weakening*. The objective in this case is not to best use the motor current to produce torque, but to best use the supply voltage to allow higher speed. The cost of speed expansion is a rapid decrease in the generated mechanical power. As shown in Figure 4.11(d), flux weakening is reflected in the generation of a negative direct current I_d that produces a demagnetizing component of armature reaction [MUL 10].

In practice, the inverter imposes phase voltages (V_a, V_b, V_c) to the motor. In turn, the rotational speed of the motor imposes the electromotive force E. Current is generated in the motor windings in response to these two quantities, and it depends on the resistances and inductances exhibited by the motor windings and by the lines connecting the inverter and the motor. The control of the inverter switches shall therefore elaborate the commutation commands s1 to s6 depending on the objective to be reached. In vector control, it shall have a precise dynamic model of the motor available as well as a precise measurement of the phase currents and of the angular position of the rotor relative to the stator.

In the torque control that minimizes windings copper losses, the inverter switches are controlled to achieve two automatic controls. One is responsible for annulling the direct component I_d. The other is responsible for producing the quadrature current I_q corresponding to the set point current I_q^*. This current is the image of the required electromagnetic torque T_e^* through the electromagnetic constant K_m of the motor. The typical functional architecture of such a vector control for PMSM is represented in Figure 4.12, in which we can observe:

– the measurement of the angular position θ of the rotor relative to the motor body and the measurement of the three currents I_a, I_b and I_c;

– the real-time calculations of Clarke and Park transforms in order to determine the intensity of the direct and quadrature components of the current (I_d and I_q, respectively) based on these measurements;

– the closed-loop control slaving the direct and transverse currents (I_d and I_q, respectively) to their set point values (0 and I_q^*, respectively). In general, this command implements a proportional and integral (PI) controller;

– the inverse Park transform that calculates in real time the three-phase voltages V_a^*, V_b^* and V_c^* to be produced based on commands elaborated by the previous controllers;

– the pulse-width modulation that elaborates the control commands of the six static switchers of the inverter from the phase voltages to be produced.

Figure 4.12 also shows the cascade type of controller that is frequently used for closed-loop position control of a load. Three nested loops can be distinguished:

– Outer position loop: the error between the position set point x^* and the effective position x is used to elaborate the motor speed set point ω_m^*.

– Motor speed middle loop: the error between the motor speed set point and the effective motor speed ω_m is used to elaborate the electromagnetic torque set point T_e^* to be generated.

– Inner current loop: field-oriented control elaborates the command of inverter switches depending on the error between the required quadrature current I_q^* and the achieved quadrature current I_q.

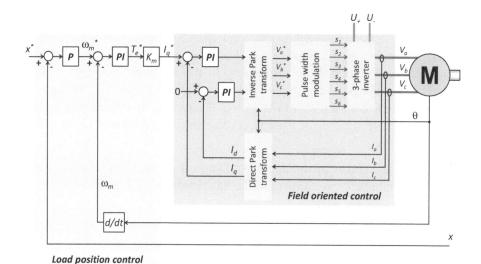

Figure 4.12. *Architecture of position control with field-oriented control of the motor*

This command structure with PI controllers is well adapted when the motor load does not generate strong coupling between the loops, which can then be adjusted in cascade, from the inner loop towards the outer loop. However, in some cases, the motor load may require other command structures or other types of controllers. This is, for example, the case of EHA, because of the hydro-mechanical mode produced by the hydrostatic loop.

Several alternatives or extensions can be applied to the architecture shown in Figure 4.12:

a) The command generally involves several protection functions (rate limiter, current, power) that are not represented in the figure. Nevertheless, it will be seen that power management is often implemented downstream of power electronics, at the motor level or even at the mechanical transmission level.

b) The architecture presented in the "vector control" block in Figure 4.12 is also applicable to asynchronous machines. In the case of these machines, the elaboration of direct and quadrature current set points takes a different approach.

c) It is possible to make without the position sensor required by field-oriented control (*sensorless control*) by using an estimator that exploits the back-electromotive voltage generated by the rotation of the rotor on the stator windings [CON 04, JOR 10]. This solution is nevertheless not accurate enough at low speed or for position control.

d) As mentioned in the beginning of section 4.3.4, it is important to isolate the power part (inverter) from the signal part (switch controls to s_1 to s_6, measurement of currents through motor windings). It is possible to simplify the line current measurement by measuring the terminal voltage across simple shunt resistors connected in series on each leg[11]. Moreover, since the machine is electrically balanced (the same effective value of currents and 120° phase difference for a three-phase machine), it is sufficient to measure two out of the three currents. The third measurement can then be used for monitoring purposes.

4.3.5. *The power architecture of a PbW actuator*

In the light of the previous considerations, the generic architecture of a PbW actuator system is presented in Figure 4.13. This shows elements in charge with power control (dynamic brake chopper and inverter) and electrical conditioning (filters, rectifier). The protection functions are not explicitly mentioned.

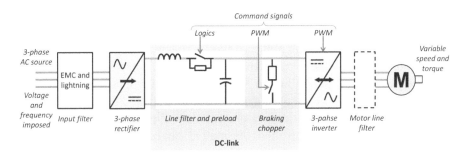

Figure 4.13. *Architecture of a PbW actuator. For a color version of this figure, see www.iste.co.uk/mare/aerospace2.zip*

---

11 But this measurement may be flawed, especially for aiding loads.

The power path from source to motor illustrated in the previous figure highlights the following elements:

– Input protection and filtering: these elements perform protection functions (lightning effects) and conditioning functions (ripples, etc.) for the power network.

– Rectifier: as already noted, motor power electronics operation relies on a DC power source. Therefore, a *DC-link* has to be locally elaborated from the three-phase source of the aircraft power network. This can be done by full-wave rectification, but this solution generates significant voltage ripples, up to six times the network frequency. The Auto-Transformer Rectifier Units (ATRU) allow the reduction of ripple factor and a doubling or tripling of the ripple frequency (12–18 ripples by network period). The Power Factor Corrector rectifiers (PFC rectifiers) produce very few ripples. In contrast to the previous ones, they must be controlled to ensure synchronicity with the network frequency, which may prove difficult for variable frequency three-phase sources such as those employed by the latest more electric aircraft [TOD 14b].

– DC bus filter: the DC bus filter has the task of eliminating the DC local bus pollution by the AC source (ripples and voltage peaks) or vice versa. It is also in charge of improving the power factor for the AC supply of the actuator. Furthermore, the preload filter function limits the strong inrush currents that may appear when the actuator is switched on, particularly those generated by downstream capacitors on the DC bus.

– *Dynamic brake chopper*: when the load driven by the motor becomes aiding, power flows from the motor to the DC bus across power electronics. In actuation applications, regenerative power is low. Therefore, the purpose is not to recover it, but rather to limit the resulting increase in DC bus voltage. To do this, the current in excess is short circuited through a braking resistor according to the principle presented in Figure 4.8(a). It is worth noting that due to its controlled dissipative characteristic, this dynamic braking chopper can also be used for the active damping of the driven load.

– Inverter: this element is in charge with the elaboration of AC voltages applied to the motor starting from the source of power supplied by the DC bus and depending on the control commands elaborated by the control electronics in order to achieve closed-loop control of the load.

Other elements may be required, such as an output filter, when there is significant distance between the inverter and the motor.

4.4. Induced, undergone or exploited effects

The previous sections of this chapter have been dedicated to the main function of power transmission and control of the mechanical load starting from a source of electric power. The implementation of this function also generates other secondary or induced physical phenomena, which are not intentionally exploited in view of the development of the said function, but result from its technological implementation. When these phenomena are undergone and alter the function, they are perceived as parasitic. On the other hand, these phenomena can be taken advantage of for the development of secondary functions, such as power management.

The effects of commutation and blocking in semiconductors, in combination with the capacitive and inductive effects that are present in the actuation chain, generate the pollution of the AC power source and can even affect its stability. Though it is of major concern in PbW [TRA 02], the network quality will not be discussed here, as it is beyond the scope of this book.

4.4.1. *Dynamics in presence*

It is of particular importance to keep in mind the orders of magnitude of various dynamics present in the actuation system. This allows us to ensure the consistency of choices made (sampling periods for the command, PWM frequency, etc.). Table 4.3 summarizes these orders of magnitude and evidences the following:

a) Thermal time constants ($\tau_t = R_t C_t$): they are related to the coupling of a thermal capacitance effect[12] C_t (the matter that stores/releases heat due to its heat capacity) and a thermal dissipation effect (the exchange of heat between matter and its environment) of equivalent thermal resistance R_t. Two types of time constants are generally identified. The rapid time constants reflect the

---

12 The term "thermal inertia" is often employed, but it is not appropriate from an energy perspective.

exchange dynamics at the heat source level: the increase in the temperature of the wires of motor windings relative to the motor body (typically several seconds) or of the substrate relative to an IGBT module baseplate (typically 0.01–0.1 s). The slow time constants reflect the dynamics of the exchange with the environment: between the motor body and the ambient environment (typically several minutes) and between the power electronics heatsink and the ambient environment (typically several seconds to several tens of seconds);

b) Electric time constants of motors ($\tau_e = L/R$): they result from the coupling of resistive effects (resistance R) and inductive effects (inductance L) of motor windings (several ms);

c) Electric period of the motor at its operating point: it is related to the rotational speed of the rotating field, for example, 833 Hz for a motor with five pairs of poles rotating at 10,000 rpm, hence a period of 1.2 ms;

d) Commutation period for a six-step control (six times the frequency of an electric cycle of the machine);

e) Period chosen for pulse-width modulation, if it relies on the use of a carrier wave (for example, typically 8–20 kHz, hence 50–125 µs);

f) Commutation time of power electronics (several 0.1 µs) and the delay adopted upon the reversal of switches on a leg (several µs);

g) Response time of the closed loops controls of position (for example, 0.25 s), speed (for example, 25 ms) and torque or current (for example, 0.5 ms);

h) Duration of motor shaft revolution relative to the body (for example, 4 ms for 15,000 rpm).

Some of these dynamics are imposed by the client requirements (for example, the response time of the position control), and others by the technological state of the art (for example, IGBT commutation time). Therefore, a small number of degrees of freedom remain available for the designer (for example, the PWM frequency) depending on the technological choices adopted for the motor (for example, the number of poles and slots).

	0.1 μs	1 μs	10 μs	100 μs	1 ms	10 ms	100 ms	1 s	10 s	100 s
Power electronics										
IGBT commutation	○									
Reversal dead time on IGBT leg			○							
PWM period (if based on carrier wave)				○						
IGBT thermal time constant							○			
Global thermal time constant									○	
Motor										
Electric time constant						○				
Electric period of the rotating field					⬭					
Trapezoidal commutation								○		
Thermal time constant of windings					⬭					
Global thermal time constant									⬭	
Output shaft revolution time					⬭					
Command										
Response time of the torque current loop						○				
Response time of the speed loop						○				
Response time of the position loop							○			

Table 4.3. Orders of magnitude of the dynamics present in the electromechanical power chain of a PbW actuator

4.4.2. *Torque ripple*

The electromagnetic torque produced by an electric motor presents ripples [HOL 96, HSU 95]. From the perspective of the actuated load, the motor also acts as a torque exciter. In most applications, the speed varies within a wide range, which includes the null value. For certain rotational speeds, the torque ripple poses the risk of exciting the eigen modes of the mechanical load driven by the motor, thus generating oscillatory instabilities that produce vibrations, noise and wear. The torque ripples have several origins:

– Cogging torque:

The cogging torque is present even in the absence of current. It results from the interaction between the magnetic fields generated by the rotor magnets and the stator slots. The variation in reluctance as a function of the position of the rotor relative to the stator generates attraction or repulsion forces that are locally similar to those generated by a spring. These forces vary periodically with the mechanical angle θ_c of the motor, the angular period of the fundamental harmonic component is determined by the least common multiple (LCM) of the number N_p of poles of the rotor and the number N_d of slots of the stator:

$$\theta_c = 2\pi / LCM(2N_p, N_d) \qquad [4.19]$$

– Fluctuation of the electromagnetic torque

The electromagnetic torque presents parasitic fluctuations caused by the non-coherence of currents (resulting from the voltage imposed by the power electronics to the stator windings) and back-electromotive voltages (resulting from the position and angular velocity of the rotor). There are two main causes for this non-coherence. The time distortion, independent of the electrical angle, is caused by the alteration in the form of current that emerges starting from the voltage elaborated by the power electronics and the strategy of control of its transistors. It can be perceived at variable moments relative to the electrical angle, for example, at the PWM period. The space harmonic distortion is caused by the imperfections in building the desired alternating flux profile in the air gap (slotted flux for BLDC motors or purely sinusoidal flux for PMSM), depending on the mechanical angle. It is clear from the basic example in Figure 4.3 that the harmonic ripple corresponds to the effect of the angle θ (equation [4.12]) and to the imperfect realization of a constant magnetic field *B*.

– Electromagnetic torque pulsations

In some motor architecture, the air gap reactances differ along the direct axis (d) and the quadrature axis (q). This difference generates electromagnetic torque pulsations that are linked to the machine structure.

4.4.3. *Energy losses*

The control and conversion of electric power into mechanical power involve relatively small energy losses. Nevertheless, they are worth being considered for two essential reasons. On the one hand, they are important for the energy consumption balance, which determines the sizing of elements located upstream in the power system, and on the other hand, they are important for the study of the thermal equilibrium of the actuator itself, which is in most cases cooled through free exchange with its environment.

4.4.3.1. *Power electronics*

From a power electronics perspective, there are conduction losses and switching losses. In what follows, only the losses generated by the inverter will be discussed. However, losses in other elements of the electronic system should be considered, such as those in rectifiers or filters, given that they are generally significant.

a) *Conduction* losses generate a voltage drop across transistors or diodes when they are in on-state (typically between 0.5 and 3 V at nominal current). This voltage drop comprises a constant part, the threshold voltage V_{c0}, and a part that varies with the current flowing through the component, equivalent to the effect of a conduction resistance R_{on}, which can be considered linear in a first approximation. The mean power dissipated by conduction is hence expressed by:

$$\mathcal{P}_{con} = V_{c0}|I| + R_{on}I^2 \qquad [4.20]$$

As mentioned in section 4.1.2.2, the silicon carbide (SiC) electronic components exhibit conduction losses reduced by a factor of 3, compared with the silicon components. They are already used in aerospace, for example, for certain free-wheel diodes on the Airbus A350.

b) *Switching* losses occur during the change of state, as the current and voltage are not instantaneously set at the value corresponding to their new operating state (null current in off-state and null voltage drop in on-state).

The loss results from the simultaneous presence of a voltage and current difference, during the transient change of state. Unfortunately, the switching losses also depend on the load connected to the power electronics, and their accurate modeling is difficult. There is a defined switching energy $\mathcal{E}_{on}$ for the turn-on and $\mathcal{E}_{off}$ for the turn-off. The data sheets of semiconductors provide these energies under reference conditions: $\mathcal{E}_{onref}$ and $\mathcal{E}_{offref}$ for supply under voltage U_{ref} and current I_{ref}. In a first approximation, these data can be scaled to effective values of voltage U and current I:

$$\mathcal{E}_{on} = \frac{U}{U_{ref}} \frac{I}{I_{ref}} \mathcal{E}_{onref} \qquad [4.21]$$

$$\mathcal{E}_{off} = \frac{U}{U_{ref}} \frac{I}{I_{ref}} \mathcal{E}_{offref} \qquad [4.22]$$

The power dissipated by switching is equal to the sum of switching energies in a unit time interval. If the switching frequency f is constant, for each semiconductor, this power can be written as

$$\mathcal{P}_{com} = f(\mathcal{E}_{on} + \mathcal{E}_{off}) \qquad [4.23]$$

Table 4.4 illustrates the orders of magnitude of the parameters and the associated losses for an integrated module 1,200 V, 75 A, having 6 IGBT with their free-wheel diodes.

Junction temperature (°C)	25	125
IGBT Threshold voltage (V)	0.5	0.6
IGBT Conduction resistance (mΩ)[*]	14	18
On-state commutation time (ns)[**]	260	290
On-state commutation energy (mJ)[**]	6.6	9.4
Off-state commutation time (ns)[**]	420	520
Off-state commutation energy (mJ)[**]	6.8	8
Diode threshold voltage (V)	0.9	0.6
Diode conduction resistance (mΩ)[*]	10	14.7
Off-state commutation time (ns)[**]		300
Off-state commutation energy (mJ)[**]		6.5

[*] Equivalent value, at 75 A
[**] Under supplier measurement conditions, 600 V and 75 A in particular

Table 4.4. *Orders of magnitude of the characteristics of an IGBT (1,200 V, 75 A) with its free-wheel diode*

NOTES.– Aerospace actuators, and especially those for flight control, operate below the maximum speed and torque most of the time (please refer to Figure 1.4 of Volume 1 for an illustration). The control efficiency of power electronics is significantly degraded under these conditions. However, it is still higher than that of the hydraulic servovalve control, which is disadvantaged by the leakage of the pilot stage, irrespective of the operating point [MAR 09].

– As shown in Table 4.4, the parameters characteristic to losses in electronic components depend on the temperature of their substrate. This dependence is far from negligible for some which can by 50% when the temperature varies by 100 °C. This is a further illustration of the strong looping of thermal effects in PbW actuators.

4.4.3.2. *Motor*

Depending on the physical area in which they occur, the losses in an electric motor can be: electrical losses, magnetic losses and mechanical losses.

a) *Electrical* losses in brushless machines originate in the resistivity of winding wires. These conduction losses (called *copper losses* in contrast with magnetic losses) are easy to calculate from the electrical resistance R of the windings and the current I across them:

$$\mathcal{P}_{con} = RI^2 \qquad [4.24]$$

b) *Magnetic* losses or *iron losses* (or still *core losses*) are caused by the rate of change in magnetic *flux density* in the magnetic circuit of the motor (the rotating field generated by the windings) or of the angular position of the rotor relative to the stator (rotating field generated by the permanent magnets). This rate of change produces a magnetic hysteresis effect and eddy currents [KRI 10].

Eddy currents appear in the metallic parts of the motor as a result of the variation in magnetic flux density to which they are subjected. The electrical resistance of the materials through which these currents flow generates a small power loss $\mathcal{P}_E$, which is proportional to the square of motor velocity and to the maximum magnetic field B_m, in a first approximation:

$$\mathcal{P}_E = k_E B_m^2 \omega_m^2 \qquad [4.25]$$

The power losses through eddy currents are thus equivalent to those that would be generated by viscous friction of coefficient $k_E B_m^2$ in the motor bearings.

The ferromagnetic materials employed in the motor exhibit the hysteresis effect in their magnetic field/induced field characteristic. When going through a cycle of this characteristic, a hysteresis loop is evidenced, and its area represents energy loss. The power loss $\mathcal{P}_h$ by magnetic hysteresis is, in a first approximation, proportional to the cycle frequency, and hence to the motor rotational speed:

$$\mathcal{P}_h = k_h B_m^\gamma |\omega_m| \qquad [4.26]$$

where the Steinmetz exponent γ ranges between 1.5 and 2.5. This appears to be equivalent to the power that would be produced by a solid friction $k_h B_m^\gamma$ in the motor bearings.

c) *Mechanical* losses are generated by friction. On the one hand, this friction is produced by the hinge bearings between the rotor and the stator. Bearing friction depends on temperature, speed and mechanical load. The detailed calculation is now well documented [SKF 03] for the steady speed regime. Unfortunately, it is not adapted to PbW actuators that operate essentially in transient regimes around the null speed. The fluid shear in the air gap between rotor and stator is a further source of friction. The fluid shear torque is generally a quadratic function of the rotational speed, with a proportionality factor k_f. The power dissipated by shear is expressed as

$$\mathcal{P}_f = k_f |\omega_m^3| \qquad [4.27]$$

In general, the loss of power in an actuator motor is mostly caused by the conduction losses directly related to the current, and therefore, to the electromagnetic torque. The magnetic hysteresis losses are small. The eddy current losses are manifest at high velocities. The fluid friction torque is generally negligible, except for when EHA have wet motors (where the rotor rotates in hydraulic fluid) at very low temperatures, as the viscosity of the hydraulic fluid is very high.

4.4.3.3. *Overall efficiency*

The plot of the power characteristic of an inverter/motor set in the mechanical power plane is highly instructive, especially if constant-power and constant-overall efficiency curves are drawn for the two operating modes (opposite load and aiding load). Figure 4.14 provides an example for a 5 kW automotive actuator (BLDC motor).

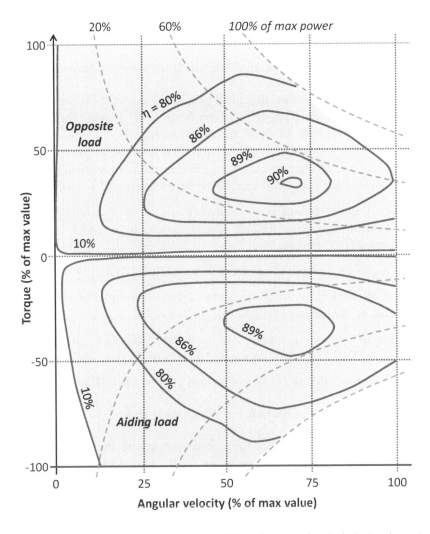

Figure 4.14. *Example of power characteristics of a motor/control electronics set (automotive application for 5 kW peak), according to [SON 16]*

It is an excellent efficiency for a wide operating range. Nevertheless, it should be recalled that, in flight control applications, actuators operate most of the time at speeds that are close to zero and at mean torques that are small compared to rated torque (0–30% depending on the actuated surface). Despite its high level at rated power point, the overall efficiency drops rapidly when approaching null speeds and torques. In Figure 4.14, it remains nevertheless above 60% at 10% of the maximum velocity and 20% of the maximum torque.

4.4.4. *Impact of concepts and architectures on performances*

The designers of electromechanical actuators have many architectural and design options at their disposal that can be used to manage the secondary effects described above or even to improve reliability: to minimize their importance if they alter the power transmission function or, on the contrary, to take advantage of them in case they can contribute to realizing secondary functions. There are many and varied degrees of freedom[13] in terms of:

– magnet assembly: surface mounted (linear or step skewed [SAH 15]), buried or with flux concentration (double layer, in V [AKI 16]), in the Halbach array [GIE 10], etc.);

– choice of the number of phases, poles and slots [DOG 11];

– type of winding (with or without overlap, distributed or concentrated, *single layer* or *double layer*) [ELR 10];

– shape of slots (full, empty, closed, bifurcated) [GIE 10], straight or skewed slots [DÖN 10], with segmented or non-segmented structure;

– association with power electronics and control mode [CAO 12].

Figure 4.15 illustrates the range of winding choices.

It is a known fact that losses due to eddy current are significantly reduced by the use of laminate and isolated plates for the parts that conduct magnetic fluxes. Moreover, some motor developments use a metal casing on the rotor (for example, to keep the magnets in place, despite centrifugal effects) or on the stator (for example, to seal the stator windings against the hydraulic fluid in a wet motor). The presence of this casing has a significant impact on the losses due to eddy current. Furthermore, it may be interesting to voluntarily increase the losses due to eddy current in order to develop load damping

---

13 As the references on this subject abound, only some of them are cited here.

functions, when the motor is not supplied (damped mode in case of failure of the electrical elements).

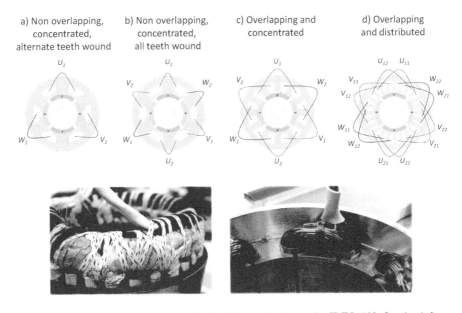

Figure 4.15. *Winding diagrams [ELR 10] and photographs [DES 12]. On the left: distributed winding, on the right: concentrated winding with single layer. For a color version of this figure, see www.iste.co.uk/mare/aerospace2.zip*

There is a reduced possibility of action on the magnetic hysteresis, because it is determined by the nature of the magnetic materials employed, of which there is limited choice. For example, the plates are made from silicon steel or grain-oriented silicon steel, which exhibit weak magnetic hysteresis and few losses due to eddy current.

There are much more options for action on the torque ripple. With the exception of time distortion, it is possible to evaluate the choices by distributed parameters numerical simulation (for example, finite elements) in electromagnetics. The results are realistic, and easy to obtain, as static simulations for various electrical angles are sufficient. An important choice resides in the form of the EMF of the motor. Associated with vector control, PMSM with sinusoidal EMF exhibits fewer pulsations than BLDC with trapezoidal EMF with their static control. Unfortunately, they are more difficult to design and develop. Their control requires accurate position measurement and strong real-time signal processing capabilities (see section 4.3.4.b). As for the cogging torque, numerous possibilities are available to the designers:

– Inverse torque ripple compensation in the torque command loop [HOL 96]. This solution can be readily implemented, as it does not affect the development of the motor. Its effectiveness is unfortunately limited by the fidelity of the electromagnetic model used for the compensation in the control rules;

– Slot *skewing* or *3-step skewed motor*. In this case the subsequent reduction of the electromagnetic constant of the motor should be accepted;

– Action on the number of poles N_p and slots N_d. It is interesting to elaborate on this degree of freedom, as it has multiple effects: on power density, maximum torque, noise, radial balancing of electromagnetic forces, flux weakening possibility, etc.

To evaluate and compare competing architectures, the number q of slots per pole and per phase is calculated for an n-phase machine:

$$q = \frac{N_d}{n N_p} \qquad [4.28]$$

The motor has *integer-slot* winding if $q = 1$, *fractional-slot* winding if $q < 1$ or *concentrated winding* if $q > 1$. The amplitude of the fundamental cogging torque appears to diminish when the least common multiple LCM ($2N_p$, N_d) increases and the highest common factor HCF ($2 N_p$, N_d) decreases. Two additional effects should be taken into consideration. First, an increase in this HCF leads to a higher imbalance of radial forces during rotation, which puts the bearings under stress and induces mechanical vibrations. Second, the commutation frequency increases with the number of poles, which increases the switching losses and magnetic hysteresis. A further important indicator is the *winding factor* k_w whose amplitude of the fundamental k_{w1} is an indicator of the effectiveness of the winding for producing electromagnetic torque, relative to a reference architecture (1 tooth per pole and per phase, same number of turns). It appears that when all the teeth are wound, k_{w1} is theoretically maximum for $q = 1/3$ and amounts to 0.955 [MEI 08].

4.4.5. Reliability

In terms of electric power systems, the typical faults that are ascribed the highest criticality[14] are: short circuits, losses of continuity, thermal runaway and magnet disconnection.

4.4.5.1. Failure rate

Beyond the adopted architectures and design concepts, reliability is strongly influenced by the type of expected response to the failure depending on the application and by technological realization and integration. It is therefore difficult to accurately provide generic values. Orders of magnitude can however be provided, such as those extracted from [CAO 12] and presented in Table 4.5.

Failure cause	Failure rate per phase and per hour of flight
Motor	
Loss of continuity	
in the windings	$1.3 \ 10^{-5}$
in the connections	$1.0 \ 10^{-6}$
other	$0.4 \ 10^{-6}$
Short circuit	
between phases	$6.7 \ 10^{-6}$
in the connections	$1.0 \ 10^{-6}$
other	$0.4 \ 10^{-6}$
Loss of power supply	$5.4 \ 10^{-5}$
Power electronics	$8.5 \ 10^{-5}$
Control	$1.3 \ 10^{-5}$
Processors	$1.0 \ 10^{-5}$

Table 4.5. *Orders of magnitude of the failure rate of a three-phase PbW actuator, according to [CAO 12]*

These orders of magnitude bring up two important points:

– the electronic part is far less reliable than the motor itself;

– the use of a simplex electrical system is not acceptable unless the criticality of its failure is minor.

---

14 See Chapter 2 "Reliability" of Volume 1.

4.4.5.2. Redundant architectures

In order to meet the reliability requirements, it is therefore necessary to design redundant and strongly segregated architectures by:

– duplicating the set electronics + motor and summing up their outlet mechanical power;

– designing fault-tolerant power electronics, for example, by adding a neutral leg or a redundant leg [ARG 08, WEL 04];

– globally designing redundant motor/power electronics.

The third option is clearly more advantageous at the level of integration in terms of mass and overall external dimensions. From this perspective, it is interesting to consider the example of the motor with 10 poles and 12 teeth ($q = 0.4$). The single-layer concentrated winding provides electrical, mechanical and thermal segregation and electromagnetic decoupling. Overall it has properties that ensure a good compromise for the actuation under mass and geometric envelope constraints. The amplitude of the fundamental winding factor is high ($k_{w1} = 0.966$) and that of its harmonics is low. For a motor with concentrated winding, there are six phases to control. Figure 4.16 presents two three-phase generic redundant architectures for this motor: modular with three independent phases or double with three phases [CAO 12]. The two solutions are redundant for the function of control and conversion of electromechanical power and require 12 transistors overall. It is also possible to control this motor as a six-phase machine and to adapt the inverter command strategy in case of failure of one of the phases [ALE 10].

4.4.5.3. Health monitoring

Electric motors and their control are well suited for Health and Usage Monitoring (HUM). The sensor signals and calculation capacity already installed for other functions can be exploited for this purpose. Beyond the simple function of use recording, HUM performances are continuously improving and it has gained in maturity in terms of fault detection (diagnostic) and fault evolution prediction (prognostic), which determine the Remaining Useful Life (RUL). As illustrated in Figure 4.17[15], these functions can be either integrated in the electronic unit associated with each actuator (case ❶) or centralized, for example, at the level of flight control computers (case ❷).

---

[15] The reader will identify the COM/MON redundant architecture presented in Chapter 2 "Reliability" of Volume 1.

Three single-phase modular redundant architecture

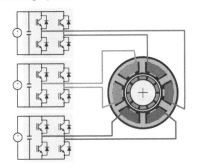

Dual 3-phase redundant architecture

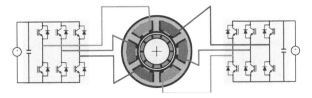

Figure 4.16. *Redundant architectures for the MPE/motor, according to [CAO 12]. For a color version of this figure, see www.iste.co.uk/mare/aerospace2.zip*

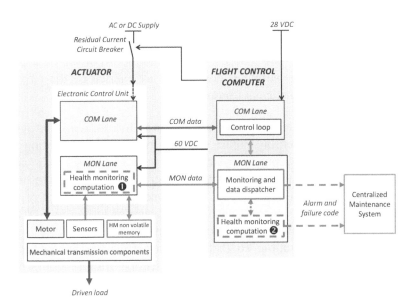

Figure 4.17. *Example of HUM functions integration, according to [TOD 14a]*

4.5. Integration

For the motor control/power electronics, it is worth considering, on the one hand, the global integration of the actuator in the aircraft and, on the other hand, the mechanical integration of electronics in its mainframe.

4.5.1. *Overall integration of the actuator*

Integration has a direct impact on the reliability of the electric power system of PbW actuators, which are subjected to very strong constraints exerted by the environment: thermal, electromagnetic and vibratory. The adopted topologies should allow the most appropriate response to the following dilemma:

– If power electronics are placed within the airframe, in a pressurized area, its environment is more favorable in terms of vibrations and electromagnetic interferences. Unfortunately, natural cooling capacity is very often insufficient, and forced cooling is required (shared cooling being possible for several electronic units). Moreover, the long power cables connecting it to the actuator motor(s) introduce parasitic capacities and inductances and they act as a source of electromagnetic interference resulting from high-frequency voltage chopping. Shielding must be strengthened and outlet filters should imperatively be integrated in the electronic modules of motor control. All these weigh down on the overall mass balance of the actuation function;

– If power electronics are placed closer to the motor, or on the actuator, the reverse happens. The strongest constraints are exerted by the vibratory, climatic and electromagnetic environments.

There are several generic power architectures for an actuation system. The first one, illustrated in Figure 4.18(a), consists of providing each actuator with the whole power system shown in Figure 4.13. It can be integrated in the actuator, as shown in Figure 4.19(b), or close to the actuator, as shown in Figure 4.19(c). The second solution consists of using the same DC source for several actuators, as illustrated in Figure 4.18(b). In this case, the rectification and filtering functions are shared. It is even possible to use regeneration, so that an actuator driving an aiding load generates DC power for the rest of the actuators. Motor control/power electronics can be integrated in one physical unit, which is often the case of the two actuators for nozzle orientation in a launcher (Figure 4.19(d)). The third solution uses one physical unit to control several actuators, as shown in

Figure 4.18(b). This allows for shared use of the inverter cooling functions, which are generally heavy and bulky. It is even possible to use only one motor control/power electronic unit for several motors, if they are used only in sequence (for example, for the extension/retraction of the landing gear).

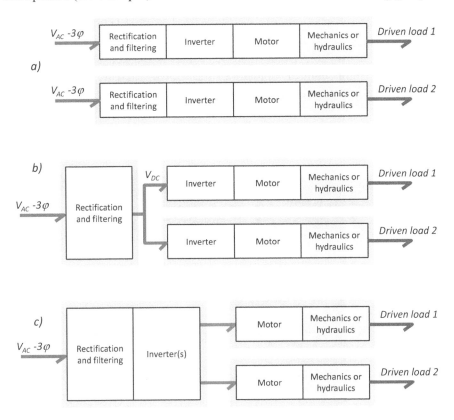

Figure 4.18. *Power architecture and integration of an actuation system*

In the case of recent Airbus (A380/A400M/A350) commercial aircraft flight control actuators, the control/power electronics of the motors are placed directly on the actuator (Figure 4.19(b)). According to [TOD 14b], the advantages in terms of mass, cooling capacity and maintainability outweigh the decrease in Mean Time Between Failure (MTBF), which is still acceptable, although reduced by a factor of 3 compared with the first option. The extent of integration is limited by geometric integration constraints or by the architecture of the actuation system. The motor control/power electronics are therefore installed closer to the actuator, which is, for example, the case for EMAs of the Boeing B787 spoiler (Figure 4.19(c)) or that of the nozzle orientation actuators of the Vega launcher (an electronic unit for two actuators, Figure 4.19(d)). The geometric envelope constraints on the flight controls of Gulfstream G650 have led to the development of a modular architecture of EBHA (actuator, HSA manifold and EHA module), in which motor control/power electronics are integrated in the electro-pump module (Figure 4.19(a)). In all these examples, it is worth noting the significant size of the motor control electronic module, which is marked by an arrow. For the Airbus flight control actuators, the power density of the motor control/power electronics integrated in its mainframe generally ranges between 1 and 2 kW/kg.

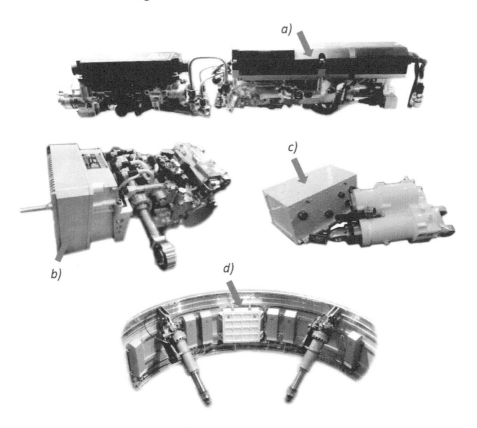

Figure 4.19. *Example of integration of the motor control electronics: a: EBHA Gulfstream G650, b: EBHA spoiler Airbus A350, c: EMA spoiler Boeing B787, d: thrust vector control of the first stage of the VEGA launcher*

4.5.2. Cooling

It is essential to keep in mind that the operating temperature of the electrical elements (electronic components, motor windings) directly impacts their service life and reliability. This is why the thermal criterion is often sizing for the actuators, which is new compared with the conventional actuators supplied by hydraulic power. The energy losses in the electrical part of PbW actuators are essentially caused by conduction resistances. The consequential heating is therefore directly linked to the history of the current supplied during the mission, or in other terms, to the electromagnetic torque generated by the motor. This is why realistic thermal sizing requires the evaluation of the temperature of the actuator elements as a function of time from a generic mission that is representative of the forces to be developed and the ambient temperature. Unfortunately, Model-Based Design (MBD) is not yet sufficiently accurate to fully replace real tests. For a given mission, a root mean square (RMS) torque T_{RMS} can be calculated, which would generate the same heating during the mission of duration t_M as the effective torque $T(t)$:

$$T_{RMS} = \sqrt{\frac{1}{t_M} \int_0^{t_M} T(t)^2 \, dt} \qquad [4.29]$$

This torque can be used for the sizing of the motor, and then of the power electronics from a thermal point of view. On the other hand, the transient torque that the motor can generate is typically 2–3 three times higher than the RMS torque. It is limited by magnetic saturation and by the demagnetizing effect of the field generated by the current on the permanent magnets. As a matter of fact, the time evolution of the torque to be supplied for a sizing mission is rarely available: the aircraft maker has never needed it for sizing actuators supplied by hydraulic power, in which energy losses depend essentially on the load speed and heat is removed by the fluid returning to the reservoir. On the other hand, the simulation of the temperature of sensitive elements (semiconductors, windings, etc.) lacks accuracy due to significant uncertainty embedded in the models of losses (mechanical, electrical, electromagnetic) and the heat exchange with the environment of the actuator. This is why sizing from a thermal point of view is a difficult task.

As shown in Figure 4.19, natural cooling often involves the use of massive heat sinks (to absorb the heat power during transient stages) that are fitted with fins (to release heat into the environment). According to [TOD 14b], the typical heat exchange capacity without forced convection is below 0.4 W/cm² of exchange surface. This is why it may be necessary to improve the release of the heat produced by actuator losses into the environment, particularly when the PbW actuators are installed in a confined environment, when they are required to develop non-null mean forces or when the idle time is short. Figure 4.20 illustrates this confinement taking an aileron EHA of Airbus A380 as an example.

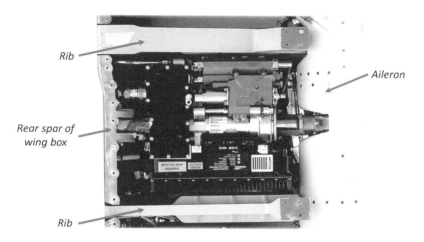

Figure 4.20. *Confinement of the aileron EHA of Airbus A380 (image according to [TOD 07])*

In this case, the common solution consists of forcing convection by facilitating the flow of air around the actuator. For example, Figure 4.21 shows the scoops installed on the underwings of the Airbus A350 with the aim of forcing convection around the EHA actuators of the aileron. A further solution consists of implementing phase change-based cooling. This is an effective method, which is well established in space technology: two-phase loops, capillary heat pipes, etc. Its use in aircraft actuation is potentially interesting, since the heat exchange capacity can reach 10 W/cm². Nevertheless, the added mass is significant, and this solution is still the subject of works aimed at optimizing and maturing it for use in aircraft [LEC 14].

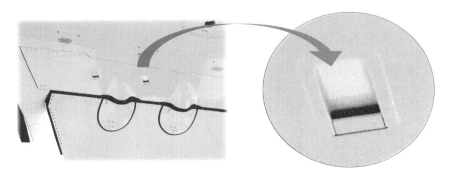

Figure 4.21. *Underwing scoops for actuator cooling (EHA of Airbus A350 aileron)*

4.5.3. *Mechanical architecture of motor control/power electronic units*

Motor control/power electronic units mechanically integrate the functions presented in Figure 4.13. They therefore comprise not only power electronics components or circuits but also sensors and control-command or signal processing electronic circuits. Consequently, the integration in a mainframe is subjected to many constraints. In terms of inside interaction, it must avoid partial discharges (see section 4.1.2.2) and protect the signal processing part against the electromagnetic aggressions generated by the power part. As for the outside, it should facilitate the evacuation of the inside generated heat, prevent the propagation of a fire starting outside the mainframe (*self-contained fire*), ensure electromagnetic compatibility and withstand the climatic and vibratory environment. In the course of developments, designers have acquired best practices that ensure the satisfaction of these requirements [TOD 07, TOD 14b, TOD 16], such as:

– permanent monitoring of the temperature of power electronics as well as the temperature of motor windings;

– controlled limitation of the maximum power drawn from the power supply network. For example, this has led to a reduction by 20% (7 kW) of the electric power required by each elevator EHA for the Airbus A380;

– employing the mainframes for mechanical, electrical and magnetic (Faraday cage) segregation between the environment, the signal part and the power part (mainframes and sensors shall be impervious and a fire-resisting partition shall separate the signal part from the power part);

– referring to the power part, the conventional electrical cables shall be replaced by *bus-bars*, minimum distances shall be provided between wires subjected to high voltages in order to prevent partial discharges and a unique shortest possible path shall be adopted for strong currents.

These good practices are illustrated by the examples in Figures 4.22 and 4.23.

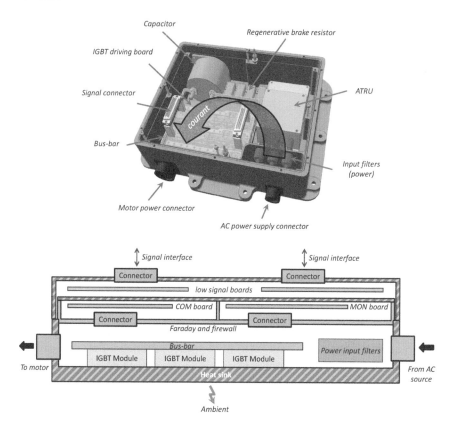

Figure 4.22. *Examples of mechanical integration of the motor control/power electronics, according to [TOD 14b] and [TOD 16]. For a color version of this figure, see www.iste.co.uk/mare/aerospace2.zip*

Electric Power Transmission and Control 137

 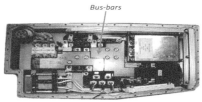

Figure 4.23. *Wiring of control/power electronics. On the left: conventional wiring; On the right: example of bus-bar use in rudder EBHA, according to [TOD 16]. For a color version of this figure, see www.iste.co.uk/mare/aerospace2.zip*

5

Electro-hydrostatic Actuators

The general concept of electro-hydrostatic actuators has been introduced in Chapter 3. Chapter 4 has focused on electric power (control electronics and motor), which is the part that PbW actuators have in common. The objective of this chapter is to provide a detailed presentation of architectures and variants of PbW actuators of EHA type, meaning actuators in which pressurized fluid is the intermediary power vector between the electric field (the power source of the actuator) and the mechanical field (the actuator-driven load). This chapter comprises three parts: historical background and maturing, solutions implemented in service and specificities of EHAs.

5.1. Historical background and maturing of EHAs

5.1.1. *PbW actuators with variable displacement pump (EHA-VD)*

PbW actuators with permanent mechanical drive allow the use of asynchronous motors; hence, they can do without power electronics. This explains why historically they were the first to be developed, at a time when semiconductors were still in their early stages. They also have a further important advantage: they generate very few high frequency electromagnetic disturbances, since they do not require strong current switching.

5.1.1.1. *Boulton Paul actuators*

As early as 1937, the company Boulton Paul Aircraft Ltd, located in Wolverhampton (UK), had already acquired good experience of using EHA-VD for airborne gun turrets. In the early 50s, the application of this type of actuator was extended to the flight controls of the English strategic bomber Avro Vulcan. There were overall 10 PFCU (Powered Flying Control Units) on each aircraft, out of which 8 were for the elevons (ailerons/elevator) and 2 for the rudder. The PFCU were supplied with three-phase voltage of 200 V at 400 Hz by four units of 40 kVA alternators. The variable displacement pump was permanently driven by the motor. The position was mechanically controlled by varying the pump displacement as a function of the error between the pilot mechanical setpoint and the actuator's rod position feedback. In case of pressure loss, a valve isolated the actuator chambers, which in a first instance blocked the rod in position. The chambers then slowly reached pressure balance thanks to internal leakages and the actuator developed null hydrostatic force. Figure 5.1 presents images of elevon actuators drawn from the advertising leaflets of the time. The advances made in terms of integration during the several years elapsed between the first version and the second ones (mk.2) are worth noting in the figure. A further important aspect on the right image is the presence of an electro-mechanical actuator for autostabilization that is integrated on the mechanical position feedback. Boulton Paul then fitted other military aircraft such as the Blackburn NA 39 (EIS 1959) or commercial aircraft such as the Viscount VC10 (EIS 1964).

Figure 5.1. *The first EH-VD actuators: example of elevons of the Avro Vulcan, on the left: Vulcan B.mk.1 (EIS 1957), on the right Vulcan B.mk.2 (EIS 1960), (images by courtesy of aviationancestry.co.uk)*

5.1.1.2. *SSAP actuator*

By the end of the 60s, the U.S. military research program known as *Survivable Flight Control System Development Program* [LOR 70] had resulted in the development of a PbW actuator (SSAP stands for Survivable Stabilator Actuator Package) for the horizontal stabilizer of an F-4 Phantom test plane [HOO 71]. This actuator was fitted with four supplies: two for electric power and two for hydraulic power. It integrated two local independent hydraulic generators that could replace centralized pressure sources. It had a tandem actuator with mechanical position feedback. The position setpoints were introduced mechanically by a quadruplex electro-mechanical actuator. The power was conventionally controlled, by means of a hydraulic metering valve supplied at constant pressure.

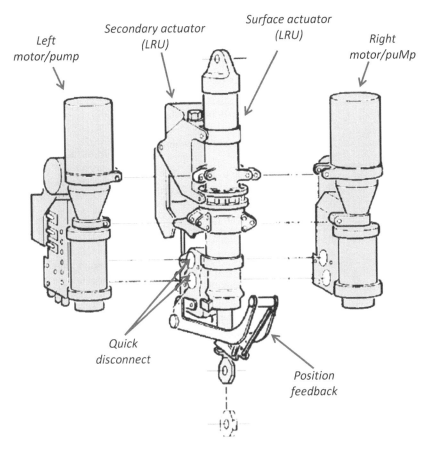

Figure 5.2. *Survivable Stabilator Actuator Package, according to [HOO 71]. For a color version of this figure, see www.iste.co.uk/mare/aerospace2.zip*

5.1.1.3. *IAP actuators*

As shown in Table 5.1, the IAP concept was first developed by Lucas in the early 80s and it continued to be investigated until 1998 [ACE 92,

ALD 93]. The objective of its development was to improve survivability of military aircraft during combat, as experience had proved that hydraulic networks were vulnerable under enemy fire. It is worth noting that in Europe this concept was evaluated by Liebherr for the aileron actuation of the Airbus A340 in the late 80s.

Date	Application	Power	Evaluation
1980–82	Auxiliary landing gear steering	Tandem ram with two independent power channels Purely mechanical control	Limited, bench test
1983–88	Auxiliary landing gear steering	Analog electric control	In-depth, bench test
1990–93	Rudder	Digital ECU with monitoring and fail-safe functions Full dual channel FbW with force equalization Active cooling by electric fan 31 kN or 11 mm/s max, 211 mm stroke, hydrostatic area of 7.3 cm² 1120 W @ 7760 rpm motor	In-depth, bench test and 5 h flight tests on Lockheed C-130
1992–98	Aileron	Fail operative/fail passive with two independent paths Force equalization by differential pressure measurement. Correction of motor power factor Communication by Mil-STD-1553 bus Max: 86.7 kN, 12.5 mm/s, 5.2 kW	1000 h flight tests beginning with 1996 on Lockheed C-141 aileron

Table 5.1. *Demonstrators of the IAP actuator concept*

The detailed power architecture of an IAP is presented in Figure 5.3. This type of actuator is exclusively supplied by electric power. As introduced in paragraph 3.4.2.2, power is controlled on demand, by varying the pump displacement, which modifies the mechanical to hydraulic power transfer ratio. Hydraulic power is then transmitted to the actuator through a hydrostatic loop, which is drawn in bold lines on the figure. Besides eliminating centralized hydraulic supply, the IAP concept preserves the hydraulics-specific advantages. On the one hand, linear actuators easily generate strong forces and low speed translational movement. On the other hand, power management functions are implemented at small mass and overall external dimensions, as the pressures in the actuator chambers reflect the mechanical force developed on the load. Figure 5.3 shows the functions implemented in conventional actuators, such as overload protection ❶, protection against outgassing or cavitation ❷ and the "hydraulic" declutching of the load ❸. However, contrary to conventional actuators, the hydraulic circuit is isolated. Additional functions shall therefore be integrated in view of local conditioning of the hydraulic fluid and displacement control, namely:

– a reservoir (or compensator or accumulator) designed to compensate the external leakages and the geometrical effects, the compressibility or dilation;

– a means for pressurizing the fluid, which shall not be too compressible, so that the actuator exhibits sufficient stiffness in response to force disturbances generated by the driven load;

– a means for power generation and control in view of actuating the mobile element of pump displacement control (typically, the swashplate of the axial piston pump, similar to the one used in pressure compensated pumps).

In the example in Figure 5.3, the displacement control actuation ❺ is assigned to a double effect cylinder associated with a servovalve supplied under constant pressure by a secondary generator ❹. For the sake of simplicity as regards the production of a maximum flow rate, this power source is constituted of a fixed displacement pump associated with a pressure relief valve. The available pressure also serves to pressurize the reservoir and to supply the refeeding function ❷. When under pressure, the pressure switch authorizes the energization of the isolation valve ❸ by a 28 VDC source that renders the actuator active to drive the load.

From a control perspective, the pump displacement control (yoke loop, Figure 3.7) can be implemented mechanically or electrically. In the first IAPs, it was purely mechanical. The displacement setpoint was elaborated by mechanical comparison of the position setpoint issued by the pilot and the actuator's rod position. It then actuated the metering valve of the yoke control function ❺. The yoke control loop was performed mechanically by using the swashplate angle as the feedback signal to drive the metering valve. An intermediate stage in the development of IAPs, called a semi-electric stage, consisted of elaborating electrically the displacement setpoint that allowed the use of an electrohydraulic servovalve. The swashplate angle feedback was still mechanical. In the more recent versions, the measurement of the swashplate angle was replaced by an electrical measurement that implemented purely electrohydraulic displacement control.

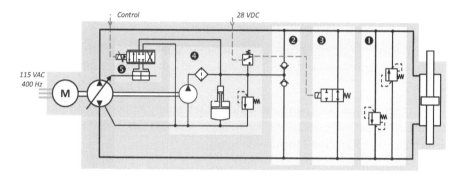

Figure 5.3. *Power architecture of an IAP actuator. For a color version of this figure, see www.iste.co.uk/mare/aerospace2.zip*

The feedback resulted from these research programs has highlighted several important points:

1) *Efficiency* is low. While power control is potentially highly efficient, since it relies on the principle of power on demand, the maximum overall efficiency of IAPs was 50%. This can be explained by the use of low efficiency asynchronous motors (80% at rated power of 1,120 W at 7,660 rpm) and pumps, whose design was a quarter of a century old. A structural cause relates to the consumption of the hydraulic generator of secondary power required for the displacement actuator.

2) *Thermal equilibrium* is difficult to predict in design, and it depends both on the actuator integration in the airframe and on the mission.

3) Proper fluid pressurizing and adapted purging procedures are required in order to reach an acceptable *actuator stiffness*, which depends directly on the presence of air or gas in the hydraulic fluid.

4) *Force equalization* was imperative for the demonstrators used for primary flight controls because, for reliability reasons, they had two power channels simultaneously active on a single load. The control structure had to reduce *force-fighting*, which was well controlled in the conventional version (identical hydraulic null for the two paths got by a common spool for the power metering valves). Several solutions have been successively introduced, such as: averaging by sending the mean value of commands to each channel, specific control of each channel depending on its pump performances and, finally, dynamic synchronization by comparing the differential pressure measurements for each piston.

5.1.1.4. EHA-VD actuators

In the early 90s, the European research program ELAC (*All Electric Flight Control Actuation*) aimed to compare the three simplex actuator technologies EHA-FD, EHA-VD and EMA for a single aisle commercial aircraft [MON 96]. Compared to the IAP concept, the EHA-VD actuator brought an innovation; the use of a torque motor for the displacement control. The *bootstrap* reservoir was replaced by a gas-pressurized piston reservoir. The hydraulic declutching was performed by a 4-orifice valve that allowed the isolation of the pump from the actuator cylinder and the interconnection of its chambers through a fixed restriction in order to generate the passive damped mode.

5.1.2. *Fixed displacement and variable speed EHA actuators*[1]

Over the past two decades, EHA-FD (fixed displacement pump driven at variable speed) have benefited from the rapid progress in the fields of power electronics, high performance electric motors and computer-assisted control. They combine the advances in the electrical field with the advantages of hydraulics in terms of power management and generation of strong forces at low speed. Compared to EHA-VD, they have better efficiency under real use conditions, which means well beyond the rated power. For example, a comparative study [BIL 98] shows that in en-route phase, the actuation of an aileron requiring most of the time 40% of the maximum force at null speed generates four times less losses in an EHA-FD than in an EHA-VD.

---

1 The author wishes to thank D. Van den Bossche, formerly in charge of the *Primary flight control actuation & hydraulics* department with *Airbus* for his suggestions after reading this paragraph.

Name of the program	EPAD	FLASH	EACS	J/IST		
Date	1990–97	1994–1996	1996–1998	1997–00		2001
Target aircraft and application	F-18 Aileron	F15 Horizontal stabilizer	F/A-18 C/D Horizontal stabilizer	F-16	F-16 Tail/Flaperon-	Launcher thrust vector control
Type		Dual tandem EHA	Dual tandem EHA	Simplex EHA	Dual tandem EHA	EHA Simplex actuator Quadruplex motorpump
Evaluation	Flight test	Ground test		900 h ground test	Flight test	Ground test
Electronics	Separated	Separated Triplex digital signal	Separated Triplex digital signal	Separated	Separated	Separated Quadruplex signal
		Moog	Moog	Moog	Parker	Moog
Stall force (daN)	59580	18681	13344		15525	33360 71100 (50 ms)
No-load speed (mm/s)	39	201	216		123/142	127[*]
Peak mechanical power (kW)		38.8	29		250	37.25
No-load power drawn (W)						
Stroke (mm)	11.4	198	180		108	>254
Pump displacement (cm^3/rev)		3.8	2.7	3.8		6.7
Rod acceleration (m/s^2)		10.16	14			
Bandwidth -3 dB (Hz)	7	5	5			6
Actuator + electronics mass (kg)	18.6 + 9				45.6 + 20.4	

[*] At 29358 daN.

Table 5.2. *Data on the development of EHA-FP for flight controls in the United States*

Electro-hydrostatic Actuators 147

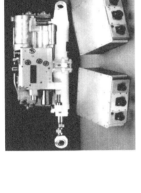

Simplex EHA (J/IST), © Moog Inc.   Horizontal stabilizer for F/A-18C/D (EACS), © Moog Inc.   Horizontal stabilator for EHA F-15, © Moog Inc.   Tail/Flaperon EHA for F-16, Parker

Thrust vector control EHA, © Moog Inc.

Figure 5.4. *Images of EHA-FP for flight controls developed in the United States*

5.1.2.1. *Development of EHA-FD for flight controls in the United States*

Similar to IAPs, the EHAs were developed in the Unites States for military applications, with the purpose of facilitating operation support and increasing aircraft survivability (Table 5.2). As shown in Figure 5.4, motor control electronics was systematically separated: on a fighter aircraft, the space available for actuator integration is very limited, all the more so when this replaces a conventional actuator. In the early 90s, a tandem aileron EHA-FD had been developed by Parker and flight-tested on HTTB C-130 [ALD 93]. Several programs have since been implemented (EPAD, EACS, FLASH, J/IST). The objective of the EPAD (Electrically Powered Actuation Device) program [NAV 97] was to demonstrate the credibility of PbW actuation for critical functions. It resulted in the development and flying of an aileron EHA-FP on an F-18. For the motor control module, it has highlighted the importance of interchangeability, open-phase detection, sensitivity of power transistors (MOS-controlled thyristors) and the manufacturing quality. Similarly, by the mid-90s, the FLASH (Fly-by-Light Advanced Systems Hardware) program, mentioned in Table 2.3 for optical flight controls, allowed Moog and Parker to develop the two dual-tandem EHA-FD actuators for the horizontal stabilizer of F-15 [ROA 97]. The ground tests confirmed overall that EHA-FD actuators are applicable to fighter aircraft in terms of bandwidth and dynamic stiffness. They showed the high impact of the pump on performances, highlighted the *pump-up* phenomenon and revealed the importance of the earliest taking into account of the electromagnetic compatibility during the design phases. Immediately afterwards, Moog also developed a horizontal stabilizer actuator for the F/A-18 C/D within the EACS (Electric Actuation and Control System) program [TRO 96], which aimed to render the PbW technology affordable. These programs were followed by the J/IST (Joint Strike Fighter Integrated Subsystems Technology) program [SMI 96] for maturing the technological solutions implemented on the Joint Strike Fighter multirole aircraft. The actuators developed by Parker within this program made possible the first flight, in October 2000, of an F-16 fighter aircraft that was modified by using exclusively PbW flight controls with EHA-FD actuators[2]. The following year, the application of EHA-FD to the orientation

---

[2] The EHA-FD control included a limitation of power capacity (force/speed) to replicate the power capacity of the conventional actuators they were replacing. In this way, the structural loads due to actuation remained unchanged. This is a further advantage of PbW, which resides in the possibility to readily restrict on demand the power characteristic of an actuator.

of a space launcher nozzle was ground tested [BAT 12]. The actuator developed by Moog included a single ram fitted with four channels for the generation and control of hydraulic power. This architecture made the actuator *double fail operative*.

5.1.2.2. *Development of EHA for flight controls in Europe*

As shown in Table 5.3, significant efforts were invested in Europe in the late 80s for the development of EHA-FD, but for different reasons. These efforts were motivated by the possibility offered by the PbW actuators of eliminating one or several networks of centralized hydraulic power on commercial aircraft. The geometric integration constraints being less significant than on weapon aircraft, it was possible to integrate motor control electronics into the actuators in order to limit electromagnetic emissions and susceptibility. Since the early 1990s, the development programs have focused on aileron actuation. This choice allows the evaluation of an actuator that is permanently loaded during a flight (the aerodynamic force on the control surface is between 20 and 40% of the maximum force that the actuator can develop) and can be more readily integrated in an existing aircraft. The actuator can be installed in the place of the backup actuator, which is supplied by the blue hydraulic circuit for the Airbus A320 family (see Figure 7.2 of Volume 1). Therefore, it is non-active in normal mode and it can be activated depending on the test phase. Similar to the EPAD program, the objective of the ELAC program [MON 96] was to evaluate through ground tests the EHA-FD, EHA-VD and EMA demonstrators. From a thermal point of view, the tests of the EHA-FD developed by Lucas were promising. However, the power drawn under stall load exceeded the previsions, presumably because of the pump internal leakages. In 1992, an EBHA developed by Lucas Air Equipment within the CVF (Commandes de Vol Futures – Future Flight Controls) program was flight tested on an Airbus A320. Following these experiments, which were the first of the type to be carried out, the patent on *EBHA and the associated system architecture* was published. In 1997, an EHA-FD developed by Liebherr within the EPICA (Electrically Powered Integrated Control Actuator) program was flight tested on an Airbus A321 [BIL 98]. The temperatures have never reached the critical value. In the course of the following programs, the stall load and the powers were increased in order to better meet the needs of the future large European carrier Airbus A380.

Name of the program	EGIDE	ELAC [MON 96]		CFV	EPICA [BIL 98]	LISA [MOO 01]		POA [DOR 07]
Date	1990-1991	1990-1992		1990-1995	1993-1997	1997-2001	1995-2000	2001-2007
Target aircraft and application	A320 Aileron			A320 Aileron	A321 Aileron	A340 Internal Aileron	Very large aircraft	Rudder A340
Type	Simplex EHA	Simplex EHA	*Simplex EHA-VD*	Simplex EBHA	Simplex EHA	Simplex EHA	Simplex EHA	EAHA
Evaluation	Ground test	Ground test	*Ground test*	Flight test (250 h)	Flight test	Flight test (130 h)		Ground test
Electronics	Separated	Integrated	*Integrated*	Integrated	Integrated	Integrated	Integrated	Integrated
Company	Samm	Lucas	*Liebherr*	Lucas	Liebherr	Lucas	Lucas	Liebherr
Stall load(daN)	4500	3382	*3300*	4700/5100	4700	15700/18800	33500	14000/18000
No-load speed(mm/s)	28	60.7	*67.6*	26 (2200 daN)	27.5	63- 10500 daN	127	29/56
Maximum motor speed (rpm)	6000	8600	*12000*		10300		10000	
Electric power (kW)								10.3
Peak mechanical power drawn (kW) Input power at stall load (kW)		0.742	*0.295*		2.08**		44.8	
Stroke (mm)	44	73.8			44.3		196	
Bandwidth (Hz)		7	*3*		2.5***			1.7/~2.5
Maximum pressure (bar)	210	207			240		255	207 (H)
Preload pressure (bar)	3-6.5				6-18			
Reservoir volume (cm^3)		220	*100*					
Mass (kg)	19	11.6			11.5		140*	

*Cylinder/manifold set and motor control electronics, **24 mm/s 3825 daN (efficiency 44%), ***with parallel actuator in damped mode

Table 5.3. *Data on the development of EHA-FP in Europe*

Electro-hydrostatic Actuators 151

Figure 5.5. *Images of EHA-FP developed in Europe*

More recently, the European activities related to EHA-FDs within the POA (Power Optimized Aircraft, 2001–2007) program were extended along two axes. The first one refers to the standardization of the electro-hydrostatic modules (EHM for Electro Hydrostatic Module), which was initiated several years ago [ARN 98]. This idea is currently being elaborated into an ARP (Aerospace Recommended Practice) document within the SAE (Society of Automotive Engineers). It stems from the interest in the modularization or standardization of the whole (motor control electronics/motor/pump) to reduce the development costs and time. An EHM can be viewed as an equivalent of the servovalve for the power control function, except that its power supply is electric instead of hydraulic. The second axis relative to EHA-FD referred to the development and ground testing of an EAHA actuator [DOR 07], as introduced in the last paragraph of section 3.4.2.4 and presented in Figure 5.6. EAHA is in fact an extension of the EBHA in which the electro-hydrostatic mode (H) and the electro-hydraulic mode (E) can be simultaneously active in the assisted mode (A), in order to increase the force and speed capacity. As shown in the upper diagram in Figure 5.6, the assisted mode is implemented by the summation at the actuator level of the flows from the two power channels. EAHA can be distinguished from the EBHA by the mode selection valve, which has three states instead of two, which imposes the use of two solenoid valves instead of one (the diagram illustrates the selector function, but its technological realization is somewhat different from this arrangement). When none of these modes are engaged, the actuator is isolated from the two sources and the chambers are by-passed through a fixed restriction to produce the damped passive mode. The other power management functions are implemented in a conventional manner (protection against overpressures, refeeding, damped passive mode, etc.). The power capacity is presented in the middle image of Figure 5.6, which shows operating domains for each mode. The electrical mode operating domain is confined within speed limits (imposed by the electronics), and power and force limits (imposed by pressure relief devices). In assisted mode, the servovalve current setpoint I_{sv} and the motor speed setpoint ω^* are elaborated according to the lower diagram in Figure 5.6. Only the electro-hydraulic mode is active when the current to be applied at the servovalve is below its rated value. Should the opposite occur, the EHM module is activated in order to supply the flow exceeding the flow available from the servovalve (❶). However, the flow supplied by the servovalve for a given current depends on the differential pressure required by the driven load[3]. This effect shall be counterbalanced by modifying the command of the EHM as a function of the signal elaborated by the measurement of the differential pressure at the level of the actuator chambers (❷). This strategy has been validated by simulation and verified by ground tests, particularly as refers to the behavior of EAHA during transient changes of mode. The frequency response in assisted mode (without load and for an amplitude of ±5 mm) has confirmed the increasing contribution of EHM starting from about 1 Hz.

5.1.2.3. *Development of EHA-FDs for landing gears*

While the landing gear actuation (doors, locks, legs extension/retraction and steering) is also a large power consumer, there are notable differences related to flight controls. At the mission time scale, the actuation function appears transient (from several seconds for the doors and legs to several minutes for the steering) or even pulsed (less than one second for the locks). The nature of functions varies: on/off for the locks, end-stop to end-stop for the doors and legs or position-controlled for steering. The actuation of landing gears requires significant power input during a very short time, and free fall in case of failure. The hydraulic power solution is well established for the actuation of landing gears, and this is one of the main obstacles encountered in the attempt to eliminate centralized hydraulic circuits. On the other hand, the need for hydraulic supply from a centralized power distribution network sets a high penalty. This is particularly true for the auxiliary landing gear, which is generally placed several tens of meters from the hydraulic center, while being close to the electrical center, near the cockpit. A first approach, implemented for the auxiliary landing gear of the Airbus A380, consisted of doing without a centralized hydraulic network and installing a backup hydraulic generator locally (see Figure 3.3 in this volume and Figure 7.8 in Volume 1).

---

3 See equation [5.16] of Volume 1.

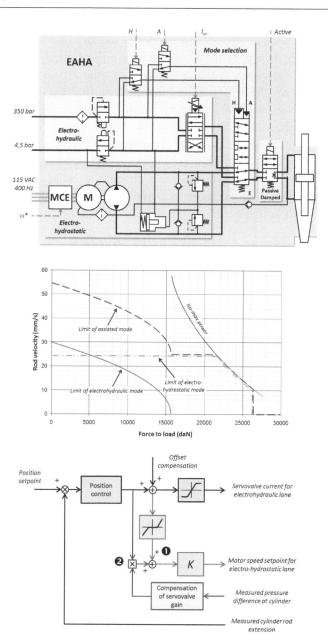

Figure 5.6. *EAHA actuator images based on [DOR 07]. Upper image: diagram, middle image: power capacity on opposing load, lower image: control strategy. For a color version of this figure, see www.iste.co.uk/mare/aerospace2*

Landing gear PbW actuation is therefore an important subject, though it enjoys less coverage than flight controls. In 1997, an evaluation of the EHA concept was carried out in Japan, for the extension and retraction of the landing gear leg of a 50 ton commercial aircraft [TAK 08]. Besides pure actuation, this function shall ensure the damping until mechanical stop in order to dissipate the kinetic energy of moving bodies and allow three operation modes: normal, free fall and on ground. Taking into account the geometric integration and the kinematics, the extension/retraction actuators are generally slender. Therefore they are naturally more intensely subjected to transversal vibrations than flight control actuators. The prototype developed and ground tested was of simplex type. Its characteristics are indicated in Table 5.4. The single rod comprised a hollow counter rod[4] that fulfilled the function of spring-pressurized reservoir. Unfortunately, this assembly configuration did not allow the easy measurement of the reservoir filling level. Moreover, this first demonstrator had not been designed under mass and reliability constraints.

Developed force (kN)	54 (extension), 107.6 (retraction)
Rod speed (mm/s)	41 (extension), 58 (retraction)
Stroke (mm)	496 mm
Motor	three-phase BLDC, 270 V, 9.33 kW, 14850 rpm
Pump	Fixed displacement, piston type, 1.8 cm^3/rev, 210 bar

Table 5.4. *Extension/retraction EHA-FD [TAK 08]*

A full auxiliary landing gear EHA-FD actuation demonstrator has been developed and tested [GRE 04] within the POA program. The key idea was to take advantage of the sequential power use for the Landing Gear Extension – Retraction System (LGERS) function. The actuators are used in sequence, for example: doors opening – unlocking – leg extension – locking. A single module for electro-hydrostatic generation has been implemented to supply the power distributed according to the adopted sequence (Figure 5.7). The free extension/free fall function was developed by specific, mechanically controlled valves.

---

4 Fixed to the body.

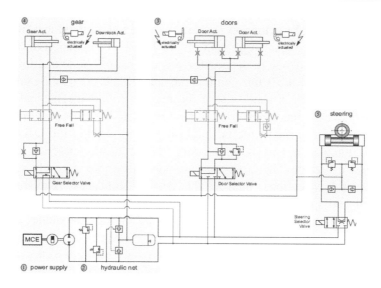

Figure 5.7. *LGERS system with EHA-FD within the POA program [GRE 04]. For a color version of this figure, see www.iste.co.uk/mare/aerospace2.zip*

Launched in 2006, the MELANY (More Electrical LANding gear sYstem) program started by exploring the landing gear steering by EHA. Figure 5.8 presents the electro-hydrostatic module and its integration on a nose landing gear of the Airbus A320.

Figure 5.8. *Landing gear steering by EHA as part of the MELANY project, extracted from [LIE 12]*

The idea of the EHM was resumed by the THERMAE II (Technology for Hydraulic-Electric Retraction Mechanism Actuation Equipment, 2011–2015) project, which brought together partners from Canada, Great Britain and Japan. The objective of the project was to prove that a LGERS application to the main landing gear of the Airbus A320 had a technology readiness level TRL5 [DAV 15, ELL 16]. Three actuators were taken into consideration: doors opening, extension/retraction and downlocking of the landing gear. The maximum forces to be developed were 300 kN in retraction mode and 50 kN for the doors, for a maximum hydraulic power of 20 kW. The locks actuation was not considered. The power architecture, Figure 5.9, implemented two channels for the generation and control of hydraulic power supplying one single actuator with a single rod from a single reservoir. One single control unit combined the functions of motor control (1 per motor, ±270 VDC supply), control of solenoid valves (28 VDC), monitoring and AFDX network communication with the avionics. There are several different technological choices for this demonstrator:

– each pump had two internal gearing stages offset by 180° and operating up to 350 bar. Due to this choice, both noise and cost were reduced, compared to an axial piston pump, which is generally used in EHAs. Geared pumps have poor performances at low speed (in the present case the efficiency is null at 1,500 rpm at 350 bar and 20°C). However, this is not a drawback for this application, since the actuators never operate to hold force close to null speed, contrary to flight controls' applications;

– the single hydraulic manifold combines the conditioning and distribution hydraulic functions. Being produced by additive manufacturing, it is easily integrated, like the two motor pumps, on the extension/retraction actuator;

– the annular cylindrical reservoir with metallic bellow is installed at the periphery of the extension/retraction cylinder, which allows for meeting the geometric envelope constraints;

– in addition to the current and motor speed loops, motor control comprises a pressure loop, which allows for making without a rod position sensor. On the one hand, this loop limits the influence of temperature on the rod speed, coming from the internal leakages of the pump. On the other hand, it serves to detect the approach of mechanical stop through the increase of pressure caused by snubbing.

This program has shown the difficulty of purging, the need to measure and monitor the filling level of the reservoir and the sensitivity of the pressure loop to transient changes of mode. Although mass and reliability balances may be negative at the level of the equipment compared to a conventional solution, the developed solution was expected to generate overall potential advantages for the aircraft.

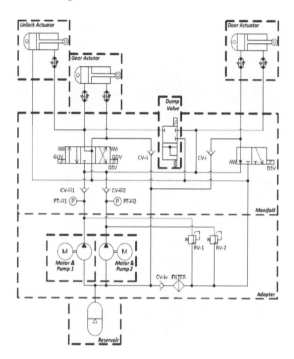

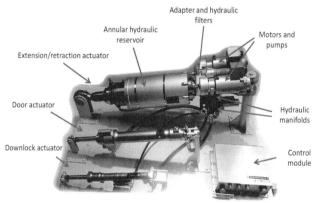

Figure 5.9. *LGERS system with EHA-FD within THERMAE II program. Upper image: power architecture [ELL 16], lower image: overall view*

5.2. EHA in service and feedback

The EHA can be perceived as a transitory solution (or rather an intermediary one) towards the *hydraulic-less* aircraft. The declutching facility, implemented in the hydraulic domain, helps prevent the jamming of a mobile surface (which is generally not allowed for primary flight controls), by switching over the actuator in free or damped mode. While waiting for mature solutions to become available for electro-mechanical actuators, particularly in order to avoid jamming, EHAs offer the possibility of eliminating a centralized hydraulic circuit in recent programs. They will therefore continue to be present in the coming decades.

Table 5.5 illustrates the use of EHAs or EBHAs for flight controls of in-service aircraft. With the exception of F-35, the PbW actuators are of simplex type and are uniquely used for backup, when a conventional actuator fails. They allow the flight in full PbW mode in case of loss of the whole set of hydraulic networks. As far as the F-35 is concerned, it uses no conventional actuator. Its EHA actuators, shown in Figure 5.10, are of simplex or dual-tandem type, depending on the surfaces. It is worth noting that the military transport aircraft Embraer KC-390, which took its first flight in early 2015, also uses EHAs and EBHAs.

	Airbus A380	Airbus A400M	Airbus A350	Gulfstream G650	Lockheed F-35
In service since:	2007	2013	2015	2012	2015
Application	Commercial	Military (transport)	Commercial	Business	Military (fighter)
Operating pressure (bar)	350	210	350	210	280
Image	Figure 3.12	Figure 3.10	Figure 4.19(b)	Figure 4.19(a)	Figure 5.10
Ailerons	8/4/0	2/2/0	6/2/0	2/0/2	0/2*/0
Elevator	4/4/0	2/2/0	2/2/0	2/0/2	0/2*/0
Rudder	0/0/4	0/0/2	2/1/0	1/0/1	0/2/0
Spoilers	12/0/4	8/0/2	8/0/4	4/0/2	
Total	24/8/8	12/4/4	18/5/4	9/0/7	0/6/0

*Dual tandem.

Table 5.5. *In-service aircraft using EHA or EBHA actuators for flight controls (the numbers indicate the type of HSA/EHA/EBHA actuator)*

Figure 5.10. *Dual-tandem flaperon EHA actuator on Lockheed F-35*

The Airbus A380 was the first series aircraft to implement EHAs. In August 2005, it used flight controls in full PbW mode for the first time, and on an extremely long route (Toulouse–Singapore). After 2.5 million hours in service, MTBF is stabilized and better than specified [TOD 16]. The operating experience feedback on the development and integration [TOD 07, VAN 06] refers essentially to the following points, some of which have already been discussed in Chapter 4:

– *Power architecture at aircraft level.* The use of PbW actuators as backup for flight controls has made it possible to use only two centralized hydraulic circuits, which resulted in a mass reduction of about 450 kg, even though an EHA is twice as heavy as a conventional actuator. 2H–2E architecture (Figure 3.3) has improved the reliability margins by increasing both the redundancy (four power sources) and their dissimilarity (hydraulic and electric). The ability to isolate a faulty PbW actuator has also improved the availability.

– *Performance of EHA pumps.* EHA pumps must exhibit repetitive volumetric efficiency in production, which should be constant throughout the aircraft life. The specificities of these pumps will be described in paragraph 5.3.1.

– *Self-contained fire.* The actuators must not propagate fire externally, particularly if they are installed near fuel tanks. This constraint has two main consequences for the design. On the one hand, the motor windings must be dry, meaning they are not soaked with hydraulic fluid, if it is used for motor cooling. The stator is therefore fluidproofed by means of a casing at the airgap level, at the cost of lowered performance. On the other hand, the electronics part must be conditioned in fire resistive units. It must be possible to isolate the actuator from the electric power source, for example by a RCCB (Residual Current Circuit Breaker, Figure 4.17) and the temperatures of the motor and of the MCE box shall be monitored.

– *Thermal equilibrium.* This aspect will be presented in detail in paragraph 5.3.4.

– *Resistance to environmental hazards.* Lightning, electromagnetic aggressions, vibrations and temperature shall be taken into account as early as possible in the integration process. The integration of motor control electronics into the actuator makes it possible to reduce the mass of filters, protections, cables and connections. It shall, however, permit its in-station replacement (LRU). In the development stage, the use of Highly Accelerated Life Testing (HALT) according to an incremental process, especially for the electronics, is particularly important. It offers the ability to evaluate margins and robustness, and also to improve reliability and maturity.

– *Limitation of the electric power demand.* In the development phase, the limitation of the maximum electric power required by each EHA allows the sizing of the electric power. The power limitation function is implemented in the actuators as software. Its final parameters are adjusted as result of flight tests and can be modified during specific phases of the flight, such as under backup electric power generation, when the available power is limited and the sequence of actuator activation in EHA mode is particularly important.

5.3. EHA specificities

EHA actuators have significant specificities that are discussed in the following paragraphs.

5.3.1. *Pumps*

EHA pumps operate under much harsher conditions than the pumps in the main hydraulic circuits. The latter have relatively high displacements, for example of 67.15 cm^3/rev for a Lockheed Martin F-22 fighter or 47 cm^3/rev for an Airbus A380. Their operation range is relatively narrow, as they run in a single power quadrant, always in the same direction of rotation, within a speed range that typically varies from 50 to 100% of the maximum speed, and they generate a quasi-constant differential pressure (210, 280 or 350 bar depending on the aircraft). They can therefore be optimized for this operation range.

On the other hand, the EHA pumps, which generally serve a single actuator, have very low displacements, ranging from 0.5 to several cm^3/rev. They must operate in all four power quadrants, with many speed reversals over a velocity range that can reach ±16,000 rpm and a differential pressure range that can reach ±350 bar. During a high amplitude step command of load position, the pump must go from quasi-null speed to maximum speed and go back to the initial speed in less than several tenths of a second. The frequency response is even more remarkable, though it does not correspond to a common operating use. It is therefore easy to understand the difficulty posed by the development of an EHA pump that shall deliver high efficiency and long service life over the whole power and temperature range. From this point of view, mechanical efficiency is particularly important, as it impacts the torque to be generated by the motor and therefore the conduction losses that are predominant in the electric and electronic parts (see section 4.4.3). The actuation of ailerons offers a good illustration of these considerations:

– during the landing phase, the driven load speeds are high and forces are moderate. The mechanical efficiency of the pump deteriorates because of the viscous losses, all the more so when fluid temperature is low;

– during the cruising phase, the forces are strong, up to 40% of the stall force, and the speeds are extremely low. The pump runs slowly, typically at 1% of its maximum speed, to generate the flow rate needed to compensate for its own internal leakages and for those of the hydraulic block. The mechanical efficiency deteriorates because the setting of fluid bearings is difficult, particularly at high temperature when the viscosity of the hydraulic fluid is very low.

The volumetric efficiency is affected by the increased clearances caused by the wear of the pump. The point of efficiency optimization of EHA pumps differs from that of the main pumps. It is not a matter of minimizing the mass of the pump for a given maximum power, but of limiting the losses for the in-service mean operating range, which have a snowball effect on the elements located upstream in the power chain.

Gear pumps have very low efficiency at low speed. Vane pumps are very silent and generate few pulses, but they are very sensitive to pollution. Radial piston pumps can run at very low speed, but they are voluminous and heavy. With the exception of particular cases (see section 5.1.2.3), the type of technology retained for EHA pumps is the axial piston pump (bent axis or inline, depending on the equipment manufacturers). It is a compact solution, which can run at high pressures and within a wide speed range.

Despite the long service life, as demonstrated by accelerated ground tests[5], the axial piston pumps of EHAs are not sufficiently mature yet to allow the use of EHAs as front-line actuators of flight controls on commercial aircraft. There are ongoing research programs that focus on extending this service life and eliminating this limitation.

5.3.2. *Filling and charging*

The EHAs involve a closed hydraulic circuit, and this has important consequences:

– any gas presence in the circuit, even the slightest, greatly deteriorates the stiffness of hydrostatic transmission, which has a direct impact on the stability of position control. The filling and air purging procedure is particularly important. Moreover, the circuit must be sufficiently pressurized to be able to avoid the risks of outgassing or vaporization/cavitation of the hydraulic fluid. From this point of view, it is interesting to increase the charge pressure of the accumulator. Unfortunately, this increase in pressure puts the seals under permanent stress. In the course of the maturation of EHAs, the charge pressure was raised from 3 bar to over 25 bar.

– when an aircraft is not active, this circuit remains pressurized by the charge accumulator. An EHA is often associated with a conventional actuator in a force summing configuration to drive a single load. When the EHA operates in passive mode, it is backdriven by the active actuator through the load they are attached to. Therefore even in this passive mode, the dynamic seals of the EHA rod are subjected to stress. Unfortunately, the technological state of the art does not allow the development of dynamic seals that are rigorously leak-free throughout the aircraft service time. One of the functions of the accumulator is to build up a fluid reserve in order to

---

5 Close to 27,000 h of flight for the PbW actuators of the Airbus A380, which is 20% of the aircraft service life.

compensate for the leakages. The sizing of such an element for the whole service life of a commercial aircraft would have a strong impact on the mass of the actuator. This is why each EHA is also connected to one of the centralized hydraulic networks through an electrical filling valve. This latter is activated on ground if the level of fluid in the accumulator is insufficient. Therefore the accumulator is essentially sized to compensate the thermal expansion of the fluid or the geometrical effects (difference between the hydrostatic areas on either side of the piston). The additional volume needed to compensate external leakage is minimal. This constraint shall be lifted in order to be able to completely eliminate the hydraulic circuits. The matter is less critical for the F-35, which does not have a centralized hydraulic circuit. The service life (in hours of flight) of a fighter aircraft is much shorter than that of a commercial aircraft, and the maintenance rates admitted per hour of flight are much more significant.

5.3.3. *Dynamic increase of mean pressure (pump-up)*

In an EHA, the mean pressure in the absence of an external load and at null speed is set at the level of pressure imposed by the accumulator. During the transient phases, the mean pressure increases as a function of speed: this is the *pump-up*. This phenomenon shall be considered, to make sure it remains within acceptable limits in terms of generated mechanical stress. It is the result of combined internal leakage of the pump, hydraulic resistance of the drain circuit towards the accumulator and hydraulic capacity of the accumulator. It is easily evidenced through simulation [MAR 06].

5.3.4. *Energy losses and thermal equilibrium*

5.3.4.1. *Efficiency*

Although power control relies on the principle of power on demand, the effective efficiency of EHAs is affected by the energy losses associated with numerous power transformations: electric/electric between the electric network and the motor, electric/mechanical for the motor, mechanical/hydraulic for the pump and hydraulic/mechanical for the cylinder. For the Airbus A380, the overall efficiency of an EHA actuator at 50% of the maximum power reaches 75% at ambient temperature and 50% at a temperature of −40°C [VAN 06]. In contrast with conventional actuators, the heat generated by the energy losses in the actuator must be dissipated (or stored) locally.

5.3.4.2. Effect of temperature

The temperature of various constituents has several effects. First, it directly impacts the service life and the reliability, which both depend on the mean operating temperature during the life of the element. Second, it directly affects the EHA efficiency and temperature, which can be illustrated by the following examples:

– the *voltage drop at rated current and the switching energy of an IGBT* increase by more than 25% when its temperature goes from 25 to 125°C;

– the *induction of a Samarium-Cobalt magnet* decreases by 4% when its temperature increases by 100°C (in normal operation zone);

– the *resistance of copper windings* increases by 42% when their temperature increases by 100°C;

– the *viscosity of the hydraulic fluid* is lowered nine times between 0 and 100°C, which increases the laminar leakage by the same factor;

– the *convective transfer coefficient* for a horizontal surface exchanging upwards with an atmosphere at 20° increases by 38% when the temperature of the surface goes from 50 to 150°C.

Only the heat transfer coefficient has a stabilizing effect on the evolution of the actuator temperature as a function of time. Other phenomena facilitate thermal divergence, as the energy losses they generate increase with their temperature. Given all these reasons, the importance of the effective operating temperature of each EHA element becomes obvious. In terms of design, thermal criteria are determinant for PbW actuators sizing. They require availability of mission profiles that are representative for the real operating conditions (including passive mode and idle time between two flights). This is a real difficulty, as these profiles shall not lead in the end to operational limitations caused by use differences between various aircraft operators. As shown in Figure 5.11, significant efforts have been carried out for more than a decade in order to obtain a numerical simulation-based prediction of the temperatures in various areas of the EHA. Unfortunately, numerical simulation is still not accurate enough for heat transfer because of the difficulty in realistically simulating the energy losses and the actuator environment. Therefore ground tests, followed by flight tests, are inescapable for accurately confirming the thermal equilibrium of EHAs.

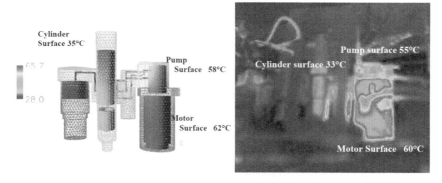

Figure 5.11. *Simulated temperature field and thermography image of a prototype EHA operating at null speed and 140 bar, according to [TAK 04]. For a color version of this figure, see www.iste.co.uk/mare/aerospace2.zip*

Figure 5.12 presents the evolution of ground-measured temperatures for a spoiler EBHA actuator of Airbus A380. For null speed and 75% of stall force (on the left), it can be noted that temperatures stabilize after 3.5 h. At 100% of the stall force, the temperatures are still not stabilized after 2.4 h, but the temperature requirement specified at 80 min is met. For these two cases, it is interesting to observe the current signal which reflects the motor speed setpoint: it is clear that the pump shall run increasingly faster in order to maintain the load pressure, as its internal leakages increase with temperature.

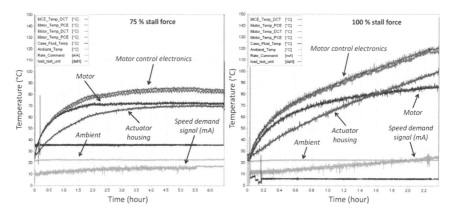

Figure 5.12. *Ground-measured temperatures for spoiler EBHA of the Airbus A380, according to [BIE 04]. For a color version of this figure, see www.iste.co.uk/mare/aerospace2.zip*

5.3.4.3. *Improvement of thermal exchanges in the EHAs*

Several means have been developed, both in terms of architectures and technological choices, to make sure that EHAs meet increasingly stricter thermal demands. With regards to architecture, it is interesting to use the hydraulic fluid in order to homogenize the temperatures within EHA in view of a better evacuation of heat from the main sources (motor) and to take advantage of a larger exchange surface. To make without a circulating pump, the drain flow rate of the pump is used to force a flow in the motor. On the other hand, this requires the isolation of the stator from the hydraulic fluid. The frequently implemented solution consists of fitting this stator with a fluidproof casing at the airgap level (*wet rotor/dry stator*), which unfortunately reduces the torque density and increases the core losses. The rotor is then flooded with fluid and no outward dynamic sealing is required. This solution generates new energy losses, by fluid shear at high speed. These losses strongly increase at low temperature, when the hydraulic fluid is very viscous, which can pose problems to the engagement of the active mode of an EHA that was operating in passive mode. Figure 5.13 shows an example of implementation of a cooling circuit. In the phase of maintaining force on a load at null speed, the pump ❶ generates a pressure on the hydraulic line A in order to develop the force required on the load. The actuator does not use any flow rate, as actuator extension does not vary. However, an internal leakage flow rate is established within the pump from the high pressure side (A) to the drain port ❷. The leakage flow then goes through a filter ❸. Then it returns to the reservoir ❹ and goes through it in order to supply line B through the refeeding valve ❺. The presence of fins that increase the convective exchange surface can be noted on the left side of the figure. These fins are also present in the examples in Figure 5.5, the black color facilitating thermal exchange by radiation.

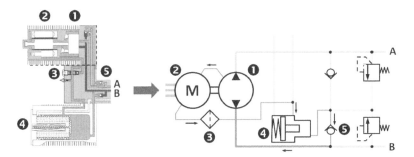

Figure 5.13. *Fluid circulation in a wet rotor/dry stator EHA-FD actuator in the phase of maintaining force on a load at null speed (drawing on the left extracted from [DOR 07]). For a color version of this figure, see www.iste.co.uk/mare/aerospace2.zip*

5.3.5. Dissymmetry

There are dissymmetric actuators that have different hydrostatic areas on either side of their piston. This choice can be imposed by the dissymmetry of forces to be generated, as in the case of spoilers that are subjected to aerodynamic forces causing the actuator to retract. It can also be imposed by the constraints of geometric integration or simply by those of assembly, as in the case of dual-tandem actuators. There are many such actuators in the LGERS. The hydrostatic area dissymmetry is problematic, since the flow rates circulating through the two lines of the hydrostatic loop are different. A first solution consists of designing an adapted architecture of the hydraulic circuit, for example by using a piloted check valve. This component allows the direct return of the flow rate in excess to the accumulator during the rod retraction phase, Figure 5.14 (top). If necessary, in the rod expansion phase, it operates as a refeeding valve. A further solution has been developed by Moog [VAL 14], which consists of fitting the valve plate of the axial piston pump with a third port and reducing the length of the kidney slot on port B, which is connected to the rod-side chamber of the cylinder (see Figure 5.14 (bottom)). Irrespective of the direction of rotation, the volume transferred to port B is then reduced, as the piston chambers are set in communication with the third port, which is connected to the accumulator. In all the cases, the accumulator must be sized to accept or return the volume displaced by the rod in the cylinder upon full retraction.

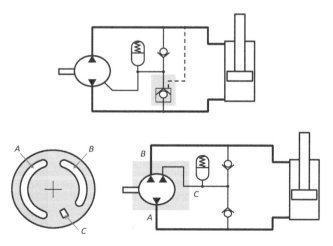

Figure 5.14. *Principles adopted for EHAs with dissymmetric actuator, according to [VAL 14]. Upper image: addition of a piloted check valve, lower image: axial piston pump with 3-port valve plate*

5.3.6. *Control*

The transmission of hydrostatic power between the pump and the actuator generates static phenomena (friction, leakages, head losses) and dynamic phenomena (essentially compressibility of hydraulic fluid). Allowing exceptions, the position control of the pump rotor in order to position the load driven by the cylinder rod is not sufficiently accurate, stable and dynamic. It is therefore imperatively necessary to measure the rod extension[6]. On the other hand, the control takes advantage of current and motor speed measurements, which are mandatory for motor control, for setting up inner loops. The transfer functions in follower mode (between the motor command and the rod position) and in regulator mode (between the disturbance force exerted by the load and the rod position) are more complex than in the case of an electro-mechanical actuator. There is a combination of electro-mechanical phenomena (motor resistance and inductance, rotor inertia) and phenomena generated by the hydrostatic transmission (leakages, head losses, friction and fluid compressibility) [ROB 05]. Depending on the predominant dynamics, the structures of the correctors of current/speed/position loops can differ from the generic structure presented in Figure 4.12. However, reaching the performances required for EHA control does not seem to encounter any real difficulties. The performances achieved are equivalent to or better than those of conventional actuators, even for weapon aircraft, which require a large bandwidth and high acceleration capability.

---

6 It is worth keeping in mind that the measurement of the load position is generally used for monitoring and diagnostic purposes, while the position sensor integrated in the actuator is used for position control purposes.

6
Electro-mechanical Actuators

The electrical part of the power chain of electromechanical actuators is similar to that of electro-hydrostatic actuators in terms of PbW power transmission for critical functions (see Chapters 3 and 4). The essential difference between them is that power transmission between motor and load is purely mechanical in EMAs and hydrostatic (eventually with a mechanical component) in EHAs. This offers interesting opportunities for designers:

– It is possible to actuate the driven load by means of an actuator that outputs a rotational movement instead of a linear one…but this involves reconsidering airframe design and actuator integration practices that date back more than 70 years.

– Maintenance is facilitated, since there are no precautions related to hydraulic fluid handling, purging or pollution to be taken…but new practices need to be acquired and disseminated, given the high voltages employed, especially in DC.

The removal of the hydraulic part brings about new constraints, each of which is a challenge to be addressed:

– It is no longer possible to take advantage of the power management functions available in the hydraulic field, such as snubbing when approaching mechanical stop, declutching and load damping for passive mode and overload protection (see Chapter 6 of Volume 1).

– It is not possible to use the hydraulic fluid as a heat carrier for cooling the electric motor and homogenizing the temperature of the actuator by taking advantage of its entire surface of exchange with the environment.

– From a load perspective, an EMA exhibits a dominant inertial effect caused by the inertia reflected by motor rotor through mechanical transmission. A hydraulic actuator exhibits a dominant elastic effect due to the compressibility of the hydraulic fluid.

– Depending on the adopted technology and reduction ratio, mechanical transmission efficiency can vary broadly as a function of load, speed, power quadrant and temperature. Transmission may even be irreversible.

– The gear backlash or the drop in transmission stiffness for weak forces transmitted can facilitate load oscillations such as *flutter* or *shimmy*.

– The service life of mechanical transmission is essentially determined by the operating conditions of sliding or rolling contacts between mobile parts (gearing system, ball or roller screws, guiding or force pick-up bearings) that were not present in hydraulic actuators (except for pumps and motors).

– Depending on the design of the actuator, the support structure and the driven load can be subjected to additional mechanical stresses, for example, in order to ensure anti-rotation in linear actuators.

– One of the generic modes of mechanical transmission failure is jamming. The jamming of the driven load is an event of high criticality in many applications such as primary flight controls or landing gears extension and steering. On the other hand, for certain applications, such as the actuation of a trim horizontal stabilizer (THS), it is the preferred fail mode, since the load is then frozen in position. For other applications, such as thrust reversers, the synchronization of actuators is the important aspect.

EMAs have been used in aerospace for many decades for *fractional horsepower* or low-criticality functions. The new opportunities provided by variable speed electric drives have only recently spurred the efforts to overcome the above-mentioned obstacles, under strong constraints imposed by mass, reliability and maintainability. The need to develop electromechanical actuation has been further stressed by the recent interest in drones, which can reach several tons in mass.

6.1. Development and operation of electromechanical actuators

As there are still a few EMAs used for high-power critical functions, operational applications will not be detailed in a separate section of this chapter. Examples of operational EMAs will be integrated in the sections focusing on different types of applications, each of which has specific requirements and constraints.

6.1.1. *Space launchers*

The evolution of EMAs for space applications is illustrated below by development (United States, Table 6.1) or commissioning (Europe, Table 6.2) examples. Space applications are generally characterized by a single mission of short duration (several minutes for a first stage of launcher), and also by bandwidths that are often highly relative to the power involved. One of the generic needs is the steering of nozzles to control the direction of the thrust vector (the module of this vector being seldom controlled). The nozzle must be steered along two axes perpendicular to its longitudinal axis in order to act on the stage yaw and pitch. The interest in electromechanical actuators related to thrust vector steering appeared very early on, but technical difficulties abounded, as shown by the example of EMAs (SPS for Service Propulsion System) of service module for the *Apollo* lunar program [MC 75, TUT 68]. A preliminary study had concluded that the electromechanical solution is overall superior to hydraulic solutions: final control and maintenance proved simpler, and yet, there was no operating feedback on hydraulic actuators used in space applications. The developed EMA was sealed and pressurized. It had two power channels in active/backup mode (DC motor, double electromagnetic clutch, parallel axes gear reducer), whose torque was summed over a single ball-screw system. During the 6-year development period (between 1962 and 1968), the supplier had to be replaced, nine different concepts or configurations had to be evaluated and speed requirements were reduced threefold, as they proved to be overrated. Finally, by 1975, no actuator failure had occurred during the 15 missions of the program.

As part of a program launched in 1974 focused on Space Shuttle mass reduction, NASA has evaluated the feasibility of replacing HSAs with EMAs in flight controls [EDG 78]. In order to be *double fail operative*, the actuator architecture had four independent chains (control, power electronics, motor), which were simultaneously active and with speed summing over a single mechanical output. Consequently, each motor was sized to generate 50% of the overall required torque and integrated a brake, in case of failure. Six-step commutation of BLDC motors has already been employed in those times. The motor current was pulse-width modulated at 700 Hz by action on two transistors connected in series and in parallel on the DC bus. The quadruplex redundancy was managed by interconnecting the four control channels and by implementing a Fault Detection Isolation and Reconfiguration (FDIR) function. Particular attention was given to energy regeneration in case of aiding load, which is an undeniable advantage of Power-On-Demand. The conclusions showed the important role played by the enhanced performances of motor magnets and power electronics, as well as higher fidelity of simulation models in design exploration. Replacing the TVC simplex, HSAs with EMAs were also approached in the field of missiles [TES 93], and it evidenced the negative effect of gear backlash on frequency response.

Between 1990 and 1993, NASA evaluated the replacement of HSAs with EMAs for steering the nozzle of the main engines of the Space Shuttle [FUL 96]. The EMA demonstrators had two electromechanical paths (three-phase induction motor and oil-lubricated parallel axes gear reducer) with torque summing over the nut of a single ball screw (Figure 6.1(a)). This time the criticality of an actuator failure during the launching phase led to operating the two paths in active/active mode. The main operating feedback concerned the problems of stability and power capacity of the power electronics that had limited the actuator tests to 37% of the predicted power of 30 kW. Around the same time [COW 93], a similar program evaluated other technological and topological choices on two types of actuators (BLDC motor, satellite roller screw, grease then dry lubrication, simplex then quadruplex motor reduction chain, 18 kW then 33.5 kW).

The example of actuation of the wing flaps of the X-38 Crew Return Vehicle (CRV) [HAG 04] is interesting in terms of the topology adopted due to the very harsh thermal environment: the temperature of the wing flap could reach 1,649°C. After several iterations, the solution retained consisted of fitting each of the two wing flaps with a rotational actuator that drove a dent-type sector with the wing flap. The actuator was *double fail operative*. Three BLDC motors operated in active mode in torque summing over a first parallel-axes spur gear reducer, followed by a cycloidal reducer, for an overall reduction ratio of 2,486.6:1. Given their maximum speed of 15,300 rpm, the motor rotors were fitted with a metallic sleeve in order to avoid jamming in case of magnet retention failure. A triplex brake helped maintain the wing flap in stow position. The overall mass of an actuator was 136 kg. Associated with each EMA were four control electronics units, distant and simultaneously active. Each electronics unit could control the three motors and communicate with the three other electronics units to ensure force equalization. The project was unfortunately abandoned in 2002, when the V201 demonstrator was 90% completed.

Electro-mechanical Actuators   175

	In service	Demonstrator	Demonstrator	Demonstrator	Demonstrator	Demonstrator
Application	Apollo TVC Service module	Space Shuttle Elevons	TVC for missile	Space Shuttle TVC Main engine		TVC
Year	1962–1975	1978	1978	1990–93		1993
Mass of actuator (kg)	9.5 to 10.5					
Continuous/peak force (kN)	3.11 / 5.78		11.1	177.9 / 444.8		
No-load speed (mm/s)	91.5 then 30.5		437	132		127
Stroke (mm)	~±35		43.7	±140		±154
Motor	DC	3-phase BLDC 270 V, 7.78 kg 12.7 kW output 13.6 Nm max 11000 rpm max	3-phase BLDC 270 VDC 9000 rpm	Induction 6 poles 3-phase 208 V, 750 Hz, 8.8 kg 26 kW DC, 52 kW peak 14700 rpm max		3-phase BLDC 270 V / 230 V 7.7 / 2.7 kg* 9300/2000 rpm max
Electronics		1 per motor 270 V, 6-step control 4x (8 Darlington)	Bipolar transistors MOSFET Driver 6-step control	12 IGBT 1200 V 150A		1 per motor 270 V 100 A 27 kW IGBT 500 V 200 A
Mechanical transmission Motoreducer Ratio Power summing Output stage Screw-nut Translational element Lubrication	Linear Duplex Torque Simplex Balls Screw	Rotational Quadruplex Speed Simplex	Balls	Linear Duplex 4:1 Torque Simplex Balls Screw Oil		Linear Duplex/Quadruplex* 9:1 Torque Simplex Roller 1 mm/rev Screw Grease / dry*
Redundant mode	Active/Standby	Active/Active		Active		4 active paths
Bandwidth (Hz)	~ 2 (−90°)	8.5 (−120°)	8 (−57°)	3.2 (−90°)		4 (−25°)

*Generation 1/Generation 2.

Table 6.1. Examples of EMA development for space applications in the United States

Certain American launchers already use EMAs for nozzle steering, for example, in the higher stage of Atlas V launchers (Figure 6.1(b)). In France, nozzle steering of the M51 strategic missile (operational since 2010) is also assigned to a direct drive EMA (Figure 6.1(c)). The motor is hollow and concentric, directly driving the nut of the inverted roller screw. In Europe, the EMAs are now used for the thrust vector control of the four stages of VEGA launcher (Table 6.2 [CAR 07, VAN 09]). The overall actuation architecture of each stage features two EMAs, for yaw and pitch, a control electronics unit in charge of the two motors (IPDU for Integrated Power Distribution Unit), lithium-ion power batteries and the electrical harness connecting these elements. Each simplex EMA comprises a PMSM motor, a parallel axes gear reducer, a nut–screw system and the control-required sensors (motor resolver, rod position and first-stage LVDT, axial force to load) and the monitoring sensors (motor temperature sensor). The three-phase sinusoidal EMF vector controlled motor is designed to operate under flux weakening within a broad range. Power requirements are thus met, while the power required by motor control electronics is limited. Roller screws are used for the first three stages, and as in the previous examples, the screw is translating. The last stage of the launcher uses a ball screw with a translating nut. Similar to all the previous examples, the motor axis is parallel to the screw–nut axis and the anti-rotation function is performed by the load and the launcher frame (see section 6.2.1.4.c). The position is controlled by the IPDU based on the instructions received by the MIL-STD-1553 bus. Having three loops (current, motor speed, rod position), the control structure is similar to the one presented in Figure 4.12. For the first stage, a force sensor is integrated into the back flange of the housing carrying the eye end of the actuator. This sensor serves for implementing an additional dynamic force feedback needed for load mode stabilization (see section 6.2.6.2). Table 6.2 shows the importance of mass and overall dimensions of the remote actuator control electronics, similar to the EHAs. Depending on stage, the IPDU mass reaches 22–128% of the mass of EMAs alone.

	P80 1st stage	Z23 2nd stage	Z9 3rd stage	AVUM 4th stage
Year	2012	2012	2012	2012
Status	In service	In service	In service	In service
Mission time (s)	120	275	395	4060
Total stroke (mm)	340	225	225	72
Pin to pin length at neutral (mm)	1048	642	642	251
Input power (kW)[*]	51	15	5	1.2
Output power (kW)	2×16	2×5.4	2×1.1	2×0.14
No-load speed (mm/s)	400	275	125	80
Force (kN)				
Blocked	100	30	20	
Pulsed	176	35	35	2.5
At maximum power	265 to 60	225 to 24	1.1 to 80	
Bandwidth (Hz)	8	8	3.5	3.5
Supply voltage (V)				
Minimum	270	275	395	75
Maximum	410	210	75	45
Mass (kg)				
Actuator alone	78	15	15	3.7
Electronics[*]	35	20.5	29.5	9.5
Batteries[*]	60	30	10	10
Electrical harness[*]	7.5	7.5	4	2
Electronics volume (dm^3)[*]	42	22	22	12.2

[*]For the 2 EMAs.

Table 6.2. *Characteristics of EMAs of the Vega European launcher*

The wing flaps actuator for IXV (Intermediate eXperimental Vehicle) had the same EMA architecture as that used on the second and third stages of VEGA (Figure 6.1(d)). The motor shaft was nevertheless fitted with a brake to keep the wing flap locked and unlock it only in the atmosphere re-entry phase [VER 11, VER 13].

a) Demonstrator of EMA for TVC [FUL 96], © Moog Inc.

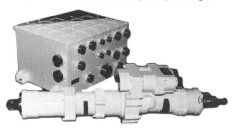
b) TVC EMA of ATLAS Centaur launcher

c) TVC EMA for strategic missile M51 (electronics from [GRA 04])

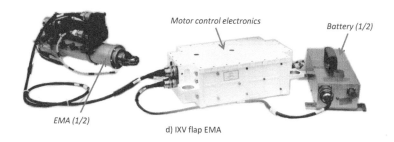
d) IXV flap EMA

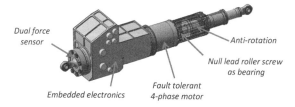
e) EMA concept for TVC with integrated electronics, from [DES 12]

Figure 6.1. *Examples of EMAs for space applications*

The ongoing developments for future programs, such as Ariane 6, are essentially focused on improving reliability (Figure 6.1(e)), for example, with fault-tolerant electronics, better inner segregation of motors, suppression of primary gear reducer or still by employing zero-pitch roller screws for the axial thrust bearings functions [DES 12].

6.1.2. *Flight controls*

6.1.2.1. *Primary flight controls*

By the mid-1980s, the EMAS (Advanced Electro Mechanical System) program had enabled the development and flight testing of a linear EMA with remote electronics for the Lockheed C-141 aileron control [NOR 86]. This actuator replaced the conventional actuator. It comprised a duplex electromechanical chain operating in active/active mode with torque summing over single ball screws. The program had shown the importance of mechanical damping in mechanical stop in case the end-stop is reached at full speed. It had also been observed that the actuator shifted to passive mode when switching the electric power supply (auxiliary power units or aircraft).

In the early 1990s, the American program EPAD focused on comparing the 3 PbW actuator technologies. It referred to the F/A-18 fighter aircraft aileron control. The EMA developed in this program [KOP 01] comprised two duplex electromechanical paths in speed summing on a single ball screw (Figure 6.2(b)). The electronic part was similar to the one used for EHAs evaluation. Each electromechanical path integrated a three-phase BLDC motor, a speed sensor and a rotor position sensor, as well as a 2:1 spur gear reducer. A further 5:1 reducer was integrated between the planetary gear speed summing unit and the nut–screw. A skewed roller clutch was integrated upstream of the summing unit. A resolver, driven by the nut through a 194:1 reducer, indirectly measured the load position. The damping of control surface in case of actuator failure was produced by short-circuiting the phases of a motor that subsequently behaved as an electromagnetic brake. This program has highlighted several points. First, the difficulties of integration in the aircraft: inside the actuator, the anti-rotation function suppressed one degree of freedom relative to the hydraulic actuator, rendering the assembly hyperstatic. Second, the importance of realistic mission profiles for the design, given that they determine the thermal equilibrium of the in-service actuator. Similar to the EMAS program, the integration in future actuators of mechanical damping capable of absorbing the kinetic energy of the motor rotor in case of end-stop at full speed proved to be imperative. The main characteristics of EMAS and EPAD are provided in Table 6.3.

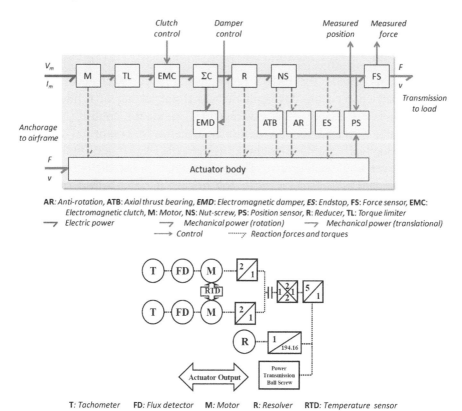

Figure 6.2. *Mechanical architectures of EMAs, Upper image: ELAC EMA, Sabca (according to [MON 96], Lower image: EPAD EMA [KOP 01]). For a color version of this figure, see www.iste.co.uk/mare/aerospace2.zip*

Project	EMAS	EPAD
Date	1985–1986	1990–1992
Aircraft and application	C-141/Aileron	F/A-18/Aileron
Flight test hours	12.5	25
Blocked force (kN)	84.7	53.65
No-load speed (mm/s)	118 (motor 9600 rpm)	214
Power point		151 mm/s 27.85 kN 4.2 kW
Stroke (mm)	138	112.8
Output backlash (mm)	0.45	0.51
Mass (kg)	29.5	12.5
Bandwidth (Hz)	4	7

Table 6.3. *Characteristics of EMAs in EMAS and EPAD projects*

The same preoccupations were present in Europe. The ELAC program, already mentioned in relation with EHAs, also started in the early 1990s and had the same objective of comparing three actuator technologies. Two EMA architectures have been developed and ground tested in this program [MON 96]. Linear simplex EMAs were used, with parallel axes reducers and remote electronics. The model examined by Lucas Air Equipment combined an epicyclical reducer (ratio 1,520:1) and a satellite roller screw (ϕ 25 mm, 5 mm lead, 0.7 μm backlash). Protection against overloads when approaching mechanical stop was provided by a torque limiter that was serially mounted on the motor shaft. The model studied by Sabca (Figure 6.2(b)) employed two spur gear reducers (overall ratio 1,491:1) and a roller screw (ϕ 32 mm, 8 mm lead, 100 μm backlash). Elastic end-stops were integrated in the nut–screw in order to limit overloads when reaching end-stop. The two models integrated a sensor of force transmitted to the load and an electromagnetic damper that acted in torque summing at clutch output. According to ELAC project conclusions, even though their efficiency was higher than that of EHAs, simplex EMAs had significant mechanical complexity and posed a serious jamming problem.

The commonly retained application for the EMA projects carried out in Europe was the aileron actuation of the Airbus A320. From 2006 to 2009, a linear direct-drive simplex EMA with remote electronics has been developed in the MOET (More Open Electrical Technologies) project. All EMA elements were concentric (motor, resolver, torque limiter, translating roller screw, rod extension sensor). Particular focus has been placed on roller screw endurance tests in order to improve the models used for service life calculations and to develop health monitoring algorithms [DEL 10]. The development and testing of motor control electronics using a two-stage matrix converter were carried out at the same time [JOM 09]. Launched in 2007, COVADIS (standing for COmmande de Vol Avec Distribution de l'Intelligence et Intégration du Système: Flight Control with Distributed Intelligence and Systems Integration) project focused on flying a linear EMA with integrated electronics and demonstrating its technology readiness level TRL6. One of the two versions developed as 114 h of flight accumulated-time [DER 11] since January 2011. Between 2011 and 2016, the Actuation 2015 project has focused on the modularization and standardization of EMA elements for cost reduction purposes [GRA 16]. It began with an analysis of the actuation needs for critical functions of aircraft (commercial, private and regional) and helicopters. This had served as a basis for defining the elements and standard interfaces (power electronics, roller screws, motors, sensors) and for demonstrating modularity: for example, motors and power electronics resulting from two industrial partnerships were interchangeable. As for motor control, standardization covered the physical modules and their interfaces (for example, PCM (Power Core Module) and CMM (Control and Monitoring Module), as well as the software and the protocols (for example, the operating system or the inverter control system).

Examples of EMA demonstrators developed for aircraft flight controls are provided in Figure 6.3. Despite all these research and development efforts, EMAs are not yet used for primary flight controls, with the exception of low-power applications, such as drones.

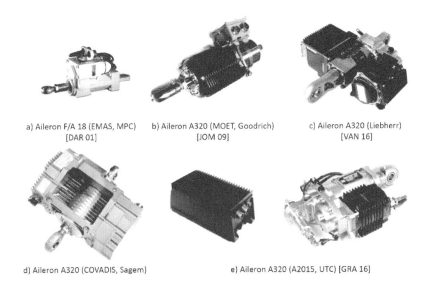

Figure 6.3. *Examples of EMA demonstrators for primary flight controls of aircraft*

6.1.2.2. *Secondary flight controls*

Jamming is often less critical for secondary flight controls. As far as the trim horizontal stabilizer is concerned, the common approach is to try freezing the load in position in case of actuator failure. It has therefore been a long time since the EMA technology has been adopted for low powers. As far as high powers are concerned, this technology is increasingly being used as second power path in recent programs, for example, for the Airbus A380 slats [BOW 04]. It has even fully replaced the hydraulic solution for the actuation of the trim horizontal stabilizer on, for example, the Embraer KC390 or on the Airbus A350. From 2001 to 2004, the DEAWS (Distributed Electrically Actuated Wing System) project focused on replacing the centralized actuation with a distributed EMA actuation [BEN 10]. In 2004, Gulfstream launched its AFC (Advanced Flight Control) program for testing FBW/PbW actuation of spoilers [WHI 07]. The first flight using EMA for the six spoilers took place at the end of 2006. As shown in Figure 6.4(a), each EMA was associated with a remote MCU (Motor Control Unit) supplied with 270 VDC generated by ATRU 18-pulse connected to three-phase 115 VAC supply. The position was controlled by REU (Remote Electronic Unit). The power chain of EMAs comprised a 12-pole PMSM, an electric off brake, a clutch and a roller screw. MCU provided an output power of 10 kW for a mass of 4 kg. REU comprised two isolated paths and had an ARINC 429 bus interface with two flight control processors, with an overall mass of 1 kg. There are also other ongoing developments related to EMA actuation of slats and wing flaps, particularly in distributed form. Finally, EMAs are already in service on the Boeing B787 for 4 out of 14 spoilers (Figure 6.4(b)).

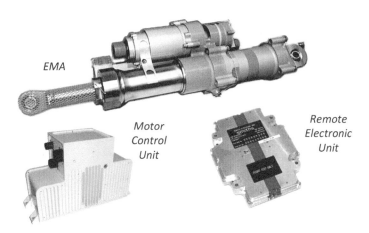

a) Spoiler EMA demonstrator (Gulfstream GV), from [WHI 07]

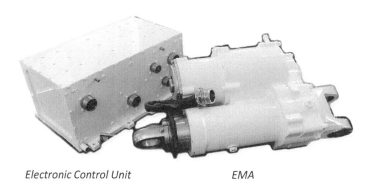

b) Spoiler EMA (Boeing B787)

Figure 6.4. *Examples of EMA for aircraft secondary flight controls*

6.1.3. *Landing gears*

6.1.3.1. *Braking system*

The brakes that act on landing gear wheels do so through friction, exerting an axial force to compress a pack of disks that are alternatively linked in rotation with the wheel and with the landing gear leg. A significant amount of kinetic energy is released as heat. It can exceed 100 MJ with an instantaneous power above 5 MW. The heat is first stored in the pack (temperature increases to 1,000° in several tens of seconds), then slowly released by exchange with neighboring bodies and ambient (below 200° after 90 min for extreme braking). Axial forces are produced by pistons distributed on the circumference of the disk pack. When the braking demand is activated, the pistons must come into contact rapidly with the pack (high-speed approach) and then control the axial force (high bandwidth modulation of force for antiskid). In parking brake mode, the pistons must maintain a minimum axial force on the pack. In the conventional solution, the force is metered by controlling the pressure applied on hydraulic pistons (see Figure 1.22).

PBW braking consists of using EMAs to apply axial force on the pack. This type of braking offers the possibility of eliminating the hydraulic fluid in the harsh environment (heat, splattering), to facilitate maintenance (LRU elements) and enhance control performances. The EABSYS (Electrically Actuated Braking System) project, launched in the late 1990s [COL 99], developed an architecture of two actuators that were isolated from the pack. The forces along EMA paths were summed and then distributed to six application points by a system of levers, flanges and pivots. Each EMA comprised a BLDC motor, a reducer and a roller screw. The overall system architecture replicated the hydraulic architecture of an axle: four EMAs (two per wheel for both right and left wheels), four remote motor control units, and two remote electronic control units. Each EMA had to develop a blocked force of 160 kN in braking mode and of 40 kN in parking mode for a mass below 20 kg. The force control bandwidth had to reach 20 Hz. In 2008, Airbus demonstrated a different concept through flight tests carried out on an A340. This time each wheel had 4 EMAs and their forces were directly applied to the pack. Similar to EABSYS, each EMA consisted of a BLDC motor, a reducer and a nut–screw system. This concept has been adopted for the electrical braking system of the Boeing B787 (EIS 2012) and Bombardier C-Series (EIS mid-2016) (Figure 6.5).

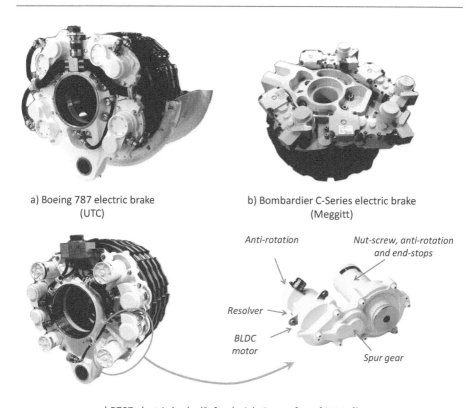

a) Boeing 787 electric brake (UTC)

b) Bombardier C-Series electric brake (Meggitt)

c) B787 electric brake (Safran), right image from [CHI 14])

d) Boeing B787 Electronic Brake Actuator Controller (Safran)

Figure 6.5. *In service EMA for PbW brake*

With its 8 braked wheels, the Boeing B787 implements 32 EMA (EBA for Electric-Brake Actuator) and 4 motor control electronic units (EBAC for Electric Brake Actuator Controller) according to the architecture presented in Figure 6.6. Each EBAC is supplied at ±135 VDC by a specific source (EBPSU for Electric Brake Power Supply Unit). Two braking processors (BSCU for Braking System Control Unit) issue the instructions for the four EBACs. According to [CHI 14], the EMA "plug and play" character provides the electric brakes with real advantages in terms of assembly and maintenance (replacement within 15 min), for usage monitoring (brake wear) and health monitoring (detection of deteriorations before loss of function) or still for operation availability and technical consistency (fault tolerance and reconfiguration).

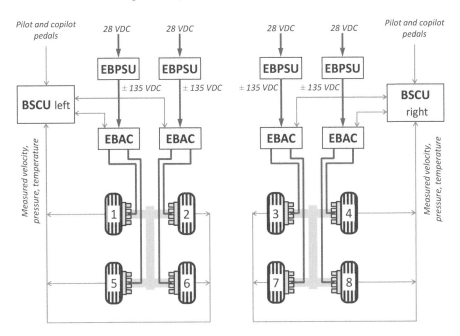

Figure 6.6. *Simplified architecture of the braking system of the Boeing B787*

6.1.3.2. Extension/retraction

The main difficulty encountered when using EMA for extension/retraction resides in the backup mode, which must allow a damped free fall of the landing gear in case of loss of the actuation function. The use of linear EMA is preferred, as it offers the possibility of preserving the overall design of the landing gear kinematics. On the other hand, in a simplex EMA, the nut–screw system downstream of the power chain constitutes an unacceptable failure node. The actuator must have properties of tolerance or resistance to jamming. A more detailed presentation of this subject will be provided in section 6.2.3. The significant reflected inertia of the motor is also difficult to slow down and damp upon mechanical stop.

Starting in mid-2000, several research programs in Europe have focused on extension/retraction actuated by EMA, particularly for the evaluation and testing of jamming-related solutions. The ELGEAR (Electric Landing Gear Extension and Retraction) project featured an EMA that consisted of the series association of two linear simplex EMAs with a parallel reducer and a motor. In the MELANY project, already cited in connection with EHAs, a direct drive EMA was used with two nested roller screws and magnetic damping [CHE 10] (Figure 6.7(a)). At the same time, the CISACS (Concept Innovant de Système d'Actionnement de Commandes de vol secondaires et de Servitude: Innovative actuation system for secondary flight controls and utilities) project explored a different path. A direct-drive EMA with a roller screw had a system of releasing the axial thrust bearing of the nut and a screw-integrated piston provided hydraulic damping [MAR 11] (Figure 6.7(b)). In the ARMLIGHT (AiRcrafts Main LandIng Gear acTuation) European project, the EMA used a secondary rod inside the screw for transmitting motion to the load. In case of nut–screw jamming, the internal rod was separated from the screw (Figure 6.7(c)).

6.1.3.3. *Steering*

EMA-based steering of landing gear also poses a problem for *free castoring*, which makes self-alignment possible in case of steering failure or during *towing*. In free castoring mode, it is also imperative to generate a damping function in order to avoid shimmy[1]. Contrary to the extension/retraction function that must move the load from one end-stop to another, steering must precisely control the angular position with a bandwidth of several Hz in pursuit, but higher in the rejection of the torque disturbances exciting the eigen modes of the landing gear leg. The power to be developed is low, but the output torque is high. An Airbus A320 requires maximum 7000 Nm, 20°/s and less than 1, 000 W. The interest in PbW is understandable, when considering the permanent power consumption of the HSA conventional solution, which is essentially linked to the servovalve leakage that can reach 700 W.

---

[1] Loss of control in active mode is also a feared event, and the use of EMA introduces no significant specificity on how this aspect is treated.

a) MELANY project EMA [CHE 10]

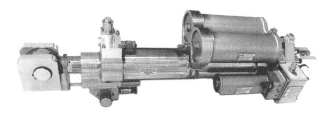

b) CISACS project EMA

c) ELGEAR project EMA (© CESA)

Figure 6.7. *Examples of EMA extension/retraction demonstrators*

In 2005, the DRESS (Distributed and Redundant Electro mechanical nose gear Steering System) project aimed to develop a PbW steering system with high reliability ($\lambda < 10^{-9}$/FH), which authorized zero visibility landing (category IIIC) for an A320 aircraft [IOR 10]. Among the evaluated power architectures, the developed and ground tested solution implemented two electromechanical paths (remote motor control electronics, PMSM motor 270 VDC, primary cycloidal reducer, electromagnetic clutch) (Figure 6.8(a)). Each path operated in active/active mode with torque summing on an output worm gear reducer. The actuator was installed at the low position of the fixed tube of the landing gear leg. The signal architecture used a local TTP bus that facilitated integration with other landing gear-dedicated systems (LGERS, braking, monitoring). It grouped 6 RDC (Remote Data Concentrator) in a star connection for the acquisitions (3 for the landing gear and 3 for the cockpit) as well as 3 CPM (Core Processing Module) for processing. The ELGEAR project launched in 2007 was also related to steering [BEN 10]. The rotational output EMA was integrated on the upper part of the fixed tube and transmitted the motion to the rotational tube, having a beneficial effect on the inertia exhibited by the leg during the retraction motion. The retained power architecture relied on two electric paths associated with a single motor. This redundant 2 × 3 phases BLDC 24-10 motor, with distributed and overlapping winding, operated at ±270 V and was PWM modulated at 10 kHz. It drove the load by means of a reducer of 595:1 ratio. An epicyclical electromagnetic clutch with double winding was inserted between the reducer output and the rotational tube, allowing the declutching of the load actuator while reducing the torque to be transmitted by the clutch. Each of the two motor control electronic units communicated with the control and monitoring processor via an ARINC 429 link. Each motor path was sized to compensate the resisting torque generated in case of short-circuit of one of the phases. The two electric paths operated in active/active mode (an active/standby mode with active path switch at each flight had been initially specified). The sampling frequency was 1 kHz for the position and speed loops and 10 kHz for the current loop. During the tests, the excessive gear clearance had limited the actuator bandwidth. After landing gear steering by EHA, the second part of the MELANY project had also been dedicated to the demonstration of steering by EMA at technological readiness level TRL4. The availability requirements being less strict than for the DRESS project, the choice was made for a simplex rotating EMA with parallel axes (Figure 6.8(b)). The power chain had a high torque motor, a Harmonic Drive reducer, a torque limiter and a spur gear. A lock was used to hold the landing gear in centered position when the actuator was idle and the aircraft was not in towing mode. The ground tests have confirmed the actuator reversibility for this mode.

a) Steering EMA (DRESS, MBD), [IOR 10b]

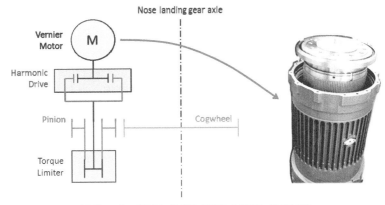

b) Steering EMA (MELANY, MBD), [LIE 12]

Figure 6.8. *Examples of EMA demonstrators for landing gear extension/retraction*

6.1.4. *Helicopters*

Helicopter primary flight controls must meet particular demands. Compared with the aircraft, their flight envelope is narrow. The minimum number of actuators is equal to the number of degrees of freedom to control (roll, pitch, yaw and climb). The mean forces (up to several 10 kN), the peak speed (up to several 100 mm/s) and the dynamic stiffness required at the blade passing frequency (typically between 10 and 30 Hz) are high. The loss of control of one actuator is only authorized for an extremely short period of time (for example, 20 ms) during which the amplitude of the uncontrolled motion shall be extremely low (for example, 5 mm). All of these constraints impose high dynamics and reliability, which are generally provided by redundant architectures. The actuators operate in active/active mode and shall at least be *single-fail functional*. It is therefore understandable why the transition to EMAs for helicopter primary flight controls is difficult.

In early 2000, the HEAT (Helicopter Electromechanical Actuation Technology) project focused on replacing the SHA flight control actuators with EMAs on Merlin EH-101 helicopters, thus passing to fully FbW and PbW flight controls [WRI 99]. This was meant to allow the removal of 2 of the 3 hydraulic circuits and of the gearbox, while it required the addition of an AC generator. The main advantages were the ease of maintenance, the best damage tolerance and a mass saving of 100 kg. The primary flight control had to be *single-fail operational*. The actuators developed (Figure 6.9(a)) had a quadruplex BLDC motor. Each motor lane was controlled by one of the four remote Motor Lane Control Units (MLCU) supplied by 270 VDC. The hollow motor directly drove a coaxial roller screw. The design was jamming-tolerant. A torque link mounted outside the actuator body ensured anti-rotation. Reliability had nevertheless not been sufficiently demonstrated during tests, and the development had to be abandoned in 2007.

At the beginning of 2010, the subject of PbW flight controls was resumed by the HEMAS (Helicopter Electro Mechanical Actuation System) European project in order to demonstrate the technology readiness level TRL5. This time flight controls had to be single mechanical fail operative and double fail operative for the rest of the system [SEE 12, ROT 14]. In a first approach, an actuator had been developed with a redundant architecture for the actuation of the swashplate (four actuators instead of three), but the pyrotechnic-operated declutch posed testability problems. In the end, a second phase retained the classical configuration with three actuators. In this latter phase, each duplex actuator summed the forces developed by two electromechanical paths (Figure 6.9(b)). Each path is itself a linear electromechanical actuator with parallel axes (roller screw, duplex PMSM motor). It is possible to disengage each path using a duplex electromechanical device (see section 6.2.3). Four actuator control electronic (ACE) units of COM/MON type are required for the two electrical paths of the motor and the two paths of activation of pyrotechnic devices. The ACE/Motor assembly is associated according to the redundant architecture presented in Figure 4.16(b). The 10-12 motor, with 2 times 3 phases, developed a peak torque of 2×4 Nm and a maximum speed of 5,500 rpm.

a) Swashplate control EMA (HEAT)

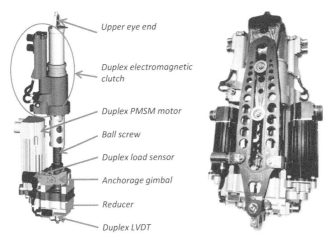

b) Swashplate control EMA (HEMAS), right image from [BIL 14]

c) EMA for IBC (Project zero), from [GIA 14]

Figure 6.9. *EMA for helicopter or convertiplane flight control. For a color version of this figure, see www.iste.co.uk/mare/aerospace2.zip*

The Individual Blade Control (IBC) is a further potential application of EMAs. It makes it possible to improve the rotor performances, to reduce the aerodynamic interactions between blades and also the level of noise and vibrations. Hydraulic power solutions tested in the 1990s consisted of integrating additional actuators in a conventional swashplate control unit. More recently, by the mid-2000s, EMAs were used for direct actuation of the blade root, without swashplate, for the 6-blade rotor of a Sikorsky CH-53G [FUE 07]. For each blade, the actuation comprised a triplex electrical chain (actuator control electronics and PMSM motor) and a 29:1 ratio concentric reducer. The whole rotated with the rotor. Triplex 270 VDC power was supplied. A 4th source was available as backup supply. The triplex motor with a mass of 23.5 kg could develop a maximum torque of 200 Nm and a maximum speed of 2,000 rpm.

The IBC was chosen for the "Project zero" unmanned convertiplane demonstrator, launched in 2010 [GIA 14]. The angle of attack of each of the three rotor blades is produced by a linear simplex direct drive EMA with remote electronics (Figure 6.9(c)). Its characteristics are presented in Table 6.3. It is worth noting that the position and speed loops use the signal provided by the resolver required by motor control, while the rod/body position is measured by a monitoring LVDT sensor.

Stroke	70 mm
Blocked force	1000 N
No-load speed	680 mm/s
Output rated power	97.3 W (330 N at 295 mm/s)
Input rated power	162 W
Mass	0.85 kg (actuator) + 0.76 kg (electronics)
Roller screw lead	2.5 mm
Reversibility	From 15% of the rated force
Bandwidth at -3 dB	3.1 kHz (current), 240 Hz (speed), 36.5 Hz (position)

Table 6.3. *Characteristics of EMA for IBC "Project zero", according to [GIA 14]*

6.1.5. *Application to engines*

Electromechanical actuation also refers to engines: actuator for Inlet Guide Vanes (IGV) steering, fuel valves, thrust reversers and other control actuators. With its Electric Thrust Reverse Actuation System (ETRAS), the Airbus A380 is a good illustration of the passage from hydraulics to PbW.

For each engine fitted with a thrust reverse, the actuator shall translate its two left and right *transcowls*, from end-stop to end-stop and imperatively in synchronicity. A simplified diagram of the power architecture is presented in Figure 6.10. The TRPU (Thrust Reverser Power Unit) supplied with 115 VAC 400 Hz essentially carries out the rectification, filtering and braking functions. The ETRAC (Electrical Thrust Reverser Actuation Controller) unit integrates the inverter with its control and electromagnetic protection systems. The PDU (Power Drive Unit) has a three-phase brushless motor with four poles, a resolver and an electromagnetic brake. Mechanical power is distributed by *flex shaft* to the two central mechanical actuators that relay it to the upper and lower mechanical actuators. Each actuator contains an angle gear box and a roller screw with a translational nut that transmits the motion to the transcowl. Secondary components ensure the functions of redundant locking and the manual command for maintenance.

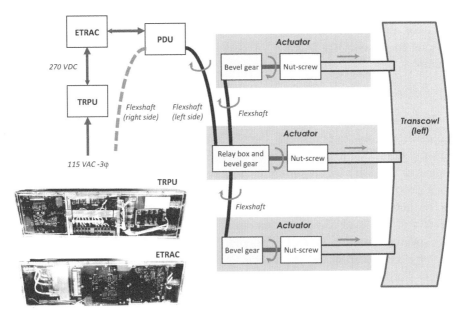

Figure 6.10. *Simplified power architecture of ETRAS of the Airbus A380. For a color version of this figure, see www.iste.co.uk/mare/aerospace2.zip*

6.2. Specificities of EMAs

As the many examples presented above indicate, after over 25 years of development, the transition to EMAs for high power and critical functions is not an easy task. The difficulty is not so much due to the electronic and electromagnetic power chain, as this is not too different from that of EHAs. It essentially resides in the realization of secondary functions (load release, locking or damping, for example) in the presence of constraints induced by exclusively mechanical power transmission (rotor reflected inertia, low reverse efficiency, etc.). The following sections will therefore concentrate on this specificity of EMAs, with a special focus on mechanical aspects (the electronic and electromechanical aspects have been addressed in Chapters 3 and 4).

6.2.1. *Power architectures*

The choice of power architecture is a complex task, which has to take into consideration many aspects: reliability, integration (mass and overall external dimensions), power capacity, natural and closed loop dynamic behavior, thermal equilibrium, etc. Designers therefore have a wide range of options to choose from, many of which have been cited in the previous examples.

6.2.1.1. Redundancy

Resorting to redundancy is often inevitable in order to meet the reliability requirements by suppressing the failure nodes. In mechanical power transmission, redundancy can be applied by force or torque summing or by (linear or angular) speed summing. Figure 2.5 provides an example for the translation of control rods, which is also applicable to the actuator's rods. As far as actuators are concerned, it is interesting to have these functions directly at the level of reducer, as shown in Figure 6.11(a). For translation, the speed summing is then realized by having two nested nut–screw systems. For rotation, an epicyclical gear is generally mounted as a speed adder. Force summing is by far the easiest to realize. Both solutions generally require the addition of one or several supplementary functions in order to obtain a fail-safe behavior for each faulty path and a fail-operational behavior for the whole transmission. In force summing, the path that cannot generate motion (jamming, for example) has to be declutched. In speed summing, the path that cannot generate force (breakage, for example) has to be blocked.

In Figure 6.11(b), three generic topologies for actuator/load association can be identified: single redundant actuator (tandem or duplex) on single load, multiple simplex actuators on single load and single simplex actuator on multiple loads. These three options can eventually be combined. The first option is frequently used for primary flight controls of combat aircraft and it is quasi-generalized on helicopters, as well as for secondary flight controls (except for spoilers) of commercial aircraft. The second is very widely used for primary flight controls of commercial aircraft. The third is used for commercial aircraft spoilers.

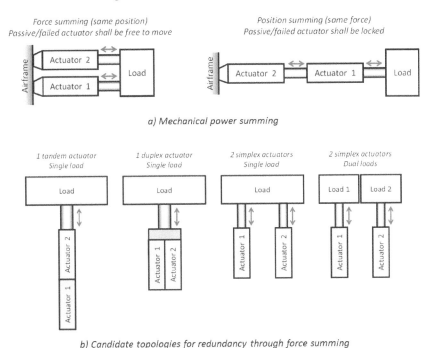

Figure 6.11. *Redundant actuation topologies. For a color version of this figure, see www.iste.co.uk/mare/aerospace2.zip*

6.2.1.2. *Outward movement*

In hydraulics, it is natural to produce a translational outward movement, which can be achieved at low costs and with good reliability by a cylinder. For several decades, the aircraft structures and the kinematics of mobile loads have been designed to take advantage of such linear actuation. According to an incremental innovation philosophy that poses few risks, most of the development programs have referred to *linear EMAs*, in an attempt to avoid challenging this know-how. Nevertheless, some programs are now exploring the option of *rotational EMAs*.

6.2.1.3. *Intermediary reducer*

Linear EMAs systematically use a nut–screw type of reducer to transform the motor rotational movement into rod translational movement. Depending on whether or not it includes an intermediary reducer between motor and nut–screw, the mechanical architecture of an EMA can be identified as *EMA gear drive* or *EMA direct drive*, respectively (Figure 6.12).

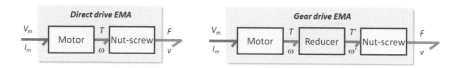

Figure 6.12. *Direct drive EMA and gear drive EMA. For a color version of this figure, see www.iste.co.uk/mare/aerospace2.zip*

The addition of an intermediary reducer and the choice of its reduction ratio result from many considerations:

a) *Mass*. As explained in section 4.2.2, the mass of a motor is closely connected with the torque it produces. In order to minimize this mass, it is therefore interesting to use a high reduction ratio. However, if an intermediary reducer is integrated, its additional mass must be taken into consideration during mass optimization. For a given load, the satellite roller screw can be made with a lower lead than the ball screw and its load capacity is higher. These two effects facilitate the removal of the reducer. However, the gain is limited by its lower mechanical efficiency, which increases the motor torque to be generated (and therefore the motor mass) to reach the force required on the load.

b) *Reflected inertia*. The inertial character of motor rotor (moment of inertia J) reflects on the load producing an effect of equivalent dead mass M_e proportional to the square of the overall transmission ratio k:

$$M_e = k^2 J \qquad [6.1]$$

$$\text{with } k = 2\pi N / p \qquad [6.2]$$

where N is the reduction ratio of the intermediary reducer and p is the lead per revolution of the nut–screw. As an example, a rotor with a moment of inertia $J = 4.5\ 10^{-5}$ kgm² with a reducer of ratio $N = 5$ and a nut–screw of lead $p = 3$ mm produces an equivalent mass of approximately 5 tons, which generally corresponds to over 10 times the equivalent translating mass of the driven load. This has several negative consequences, which have been evidenced during various development programs. The first concerns the kinetic energy E stored by the rotor, which has to be absorbed during the arrival to end-stop at high angular speed ω:

$$E = \tfrac{1}{2} J \omega^2$$

Considering the previous values, this energy amounts to $E = 24.7$ J when the motor shaft rotates at 10,000 rpm, which corresponds to the kinetic energy of a mass of 5 tons moving at 100 mm/s. In the absence of a specific device for absorbing this energy, the impact upon arrival to end-stop is much more substantial in mechanical resistance than the forces specified for covering the actuation need. The second negative effect relates to the protection against transient overloads (for example, a gust on a flight control surface). The reflected inertia does not allow the actuator to rapidly fade away in front of the external disruptive force. It is therefore necessary to provide a torque limiting function. Very often, the bandwidth of the position control or the available power is not sufficiently high to have this function performed actively through control. A mechanical function limiting the force shall therefore be integrated in the actuator, closest to the source of disturbance. A third negative effect refers to the combined influence of reflected inertia and transmission compliance on the natural dynamics of the actuator (see section 6.2.6).

c) *Backlash and compliance*. Except for particular dispositions, standard mechanical reducers have a functional backlash (typically several minutes of angle or several micrometers for the best). This clearance is harmful for the service life of the contacts and for control stability when the actuators have to operate close to null transmitted force (for example, a rudder control surface). Preloading allows for suppressing the backlash, but to the detriment of efficiency and service life, as the two faces of teeth or threads of the reducer are simultaneously in contact and loaded. While suppressing the backlash, loading gives rise to a *lost motion* region, as shown in Figure 6.13.

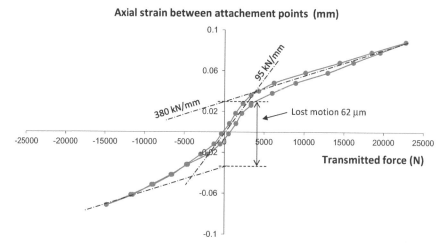

Figure 6.13. *Overall mechanical stiffness of an EMA prototype (50 kN/100 mm/s), according to [KAR 09]. For a color version of this figure, see www.iste.co.uk/mare/aerospace2.zip*

6.2.1.4. *Internal integration of the actuator*

a) Motor layout in relation to the nut–screw. Several topologies of EMA internal integration can be distinguished, depending on how the motor is fitted in relation to the nut–screw, as shown in Figure 6.14: in line, concentric, with perpendicular axes or still with parallel axes. The last two solutions impose the use of an intermediary reducer. Few developments have retained the third solution, which requires conical gears that are overall more constraining than the spur gears. As they reduce the total length of the actuator, the concentric and parallel axes solutions are well adapted to aircraft primary flight controls, as currently integrated in the airframe.

b) Nut–screw layout. If the nut in a nut–screw system is defined as a female element and the screw as a male element, four construction layouts can be identified, as shown in Figure 6.15, depending on whether the nut or the screw are in translational movement and all the threads of the nut (conventional solution) or of the screw (reverse screw) are in contact at each moment. The choice of configuration has an impact on the functions of guiding and external sealing at the rod/body level and also on the integration possibilities of the motor and gear assembly.

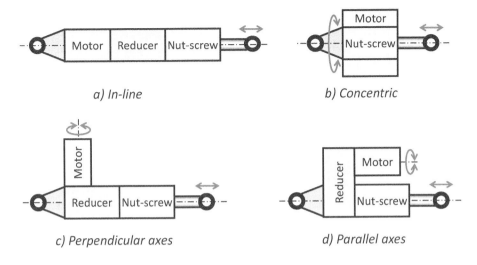

Figure 6.14. *Motor/nut–screw integration in an EMA. For a color version of this figure, see www.iste.co.uk/mare/aerospace2.zip*

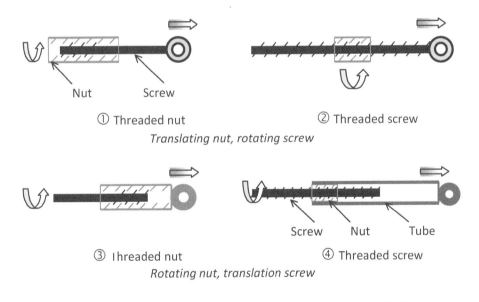

Figure 6.15. *Nut–screw layout in an EMA, according to [KAR 06]. For a color version of this figure, see www.iste.co.uk/mare/aerospace2.zip*

c) Anti-rotation. The realization of the anti-rotation function is also worth considering here. In a nut–screw system, the part that has to functionally rotate must be blocked in translation and the part that has to translate must be blocked in rotation. The axial reaction force generated by the nut–screw on its rotating part is equivalent to the pressure force exerted on the body in the chambers of a hydraulic actuator. The new element is the reaction torque pick-up. Indeed, in an HSA, there is no need to block the rod regarding its rotation relative to the body. This mobility can even contribute to a certain extent to the isostatic integration of the actuator in the aircraft (see EPAD program). The situation is different for an EMA. The reaction torques shall therefore be countered in nut–screw systems, motors (reaction of rotor on the stator) and reducers[2]. The anti-rotation function that picks up these torques can be realized at the actuator level, for example, by a torque link outside the body (Figure 6.9(a)) or still by a slide linkage inside the body. A further choice is to realize the anti-rotation function outside the actuator along the force path EMA rod – driven load – actuator support structure – EMA body. This solution couples the sizing of the load, the structure, and the actuator, as reaction torques depend on its internal design besides the operational forces to be exerted on the load. Furthermore, it requires the angular blocking of eye ends of actuator in order to realize a gimbal joint, which can be noted in Figure 6.16 for the VEGA launcher first-stage EMA.

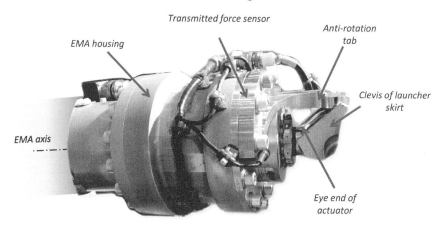

Figure 6.16. *Realization of the anti-rotation function in an EMA*

---

2 For example, in a reducer with a ratio of 5, 4/5 or 6/5 (depending on layout), the output torque has to be picked up by the body to ensure the mechanical equilibrium of the reducer.

d) Realization of secondary functions by epicyclical gear. In power transmission, epicyclical gears (Figure 6.17) are of interest. Depending on the motion allowed for the planet gears and planet carrier relative to the casing, they offer the possibility to compactly realize several types of functions:

– speed summing to associate two electromechanical power paths;

– in-line reducer (coaxial input and output shafts) with high reduction ratio, particularly by geometric nesting of several epicyclical gears that act in series in the power chain;

– torque limiter or compact clutch by installing friction disks between the inner and outer planet gears (similar to the ELGEAR program) or by blocking the outer planet gear relative to the casing (similar to the ELAC program, Lucas version). The interest is that the brake or the disks pick up only a fraction of the torque transmitted to the load.

The Harmonic Drive reducers, used for the EMA in the MELANY program, are a particularly compact realization of an epicyclical reducer. Unfortunately, the epicyclical gears involve many gears and require several bearings for guiding various bodies, all of which are potential sources of jamming.

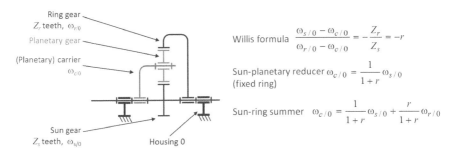

Figure 6.17. *Possibilities for using the plane epicyclical gear. For a color version of this figure, see www.iste.co.uk/mare/aerospace2.zip*

6.2.2. *Power management functions*

It is obviously tempting to try to build control-based power management functions that are highly flexible and easy to implement. As will be shown below, these possibilities can only be exploited partially due to the imperfections of the mechanical power chain (inertia, friction, etc.) and can no longer be available in case of failure of one of the control and power chain elements. Therefore, the functions are most often realized in the mechanical domain, the closest to the need. However, for mass and overall external dimensions purposes, they are integrated in the EMA in the rotational domain upstream of the nut–screw, even upstream of the intermediary reducer, if it exists. The needs and the corresponding solutions, as evidenced in the previous sections, refer to:

– *Declutching* or *locking* in order to render the failed path fail-safe in a redundant architecture with force or speed summing. The brakes and the electromagnetic clutches with friction disks are good options, since they can be activated under movement, which is not allowed by teeth solutions.

– *Damping at mechanical stop*, to limit the forces upon arrival in endstop at full speed. With the exception of LGER functions, the dominant inertial effect is produced by the rotor of the motor. The objective is to dissipate kinetic energy the closest to the rotor, without passing the forces through reducers (gear teeth, balls or rollers of nut–screw systems). It is tempting to use control in order to generate deceleration when approaching the end-stop, so that kinetic energy upon impact is reduced. This solution is often implemented for landing gear extension/retraction applications or for thrust reverser deployment/stowing applications. It nevertheless requires detecting the approach of end-stop and it is not achieved in case of actuator failure. Therefore, it is possible to use a torque limiter at motor shaft output, as, for example, the SABCA version of the ELAC project (see section 6.1.2.1). Other solutions use elastic elements or dampers. A frequently retained solution, particularly for THS actuators, consists of using a dog-teeth device design that generates a rotating obstacle engaged by the position of the translational element of the nut–screw, as illustrated in Figure 6.18.

– *Damping*. Except for functions of actuation from end-stop to end-stop, damping the load is generally required when the actuator is in passive mode, for example, to prevent flutter or shimmy. A first approach consists of taking advantage of the natural dissipative character of the elements of the actuator: friction in mechanical transmission under aiding loads and core losses in case the motor has no power supply or is open-circuited. Transmission damping by friction is rather undergone, but it may be sufficient. It is difficult to accurately predict it, as it depends strongly on the magnitude of force, preloading, speed and temperature. Damping by the motor requires the rotor to rotate, which goes against a good dynamic stiffness of the EMA, because of the inertia of the rotor, as mentioned previously. The motor core losses can be taken advantage of (see section 4.3.2) at null current in the windings, which can be controlled by design. Furthermore, the motor windings can be short-circuited, in order to have the motor operate as an electromagnetic brake. An example of a mechanical characteristic obtained in these conditions is presented in Figure 6.19. This solution has a significant impact on thermal motor sizing. In order to avoid rotating the motor, the electromagnetic brake function can be realized by a dedicated element, as is the case for the ELAC actuator developed by Lucas (see section 6.1.2.1).

– *Overload protection*. It is obvious that force limiting can be controlled by introducing saturation in the quadrature current loop. As mentioned previously, it only partially reflects the force transmitted to the load because of torques related to inertia (proportional to inertia and acceleration) and to friction effects (functions of speed, force and temperature). This is why it may be necessary to introduce a torque limiting function closest to the load.

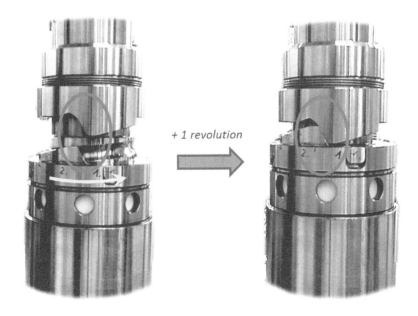

Figure 6.18. *Mechanical end-stop by dog-teeth device. For a color version of this figure, see www.iste.co.uk/mare/aerospace2.zip*

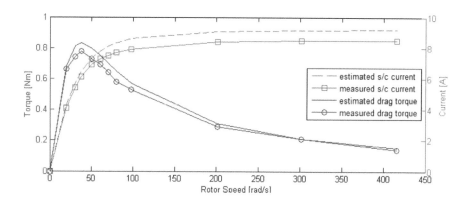

Figure 6.19. *Typical characteristic of a short-circuited permanent magnet motor, according to [ROT 14]. For a color version of this figure, see www.iste.co.uk/mare/aerospace2.zip*

6.2.3. *Jamming*

Mechanical losses in EMA are essentially caused by the friction between mechanical parts that are mobile relative to one another. In order to reduce friction losses, slip is replaced by rolling by inserting balls or rollers between the two bodies in relative motion (for example, bearing, nut–screw, anti-rotation slide, etc.). The friction coefficient typically[3] varies between 0.1 and 0.2 for an ACME screw, between 0.01 and 0.04 for a roller screw and between 0.001 and 0.006 for a ball screw. Experience indicates that besides the power transformation elements (gear reducers, nut–screws), guidings and bearings, when they are loaded, have a non-negligible contribution to friction losses (for example, at rated force, the friction produced by an axial thrust bearing can amount to over 30% of the friction produced by the nut–screw it is associated with). In the end, the overall efficiency of the mechanical transmission is reduced by all these loss sources, to which we can add the losses due to lubricant splashing or shearing and to the dynamic seals. These losses increase very rapidly when temperature drops to lower limits. Figure 6.20 illustrates the overall mechanical efficiency measured on an EMA prototype at ambient temperature. It is worth noting the irreversibility for an aiding load below 3 kN (which corresponds to the preload value to avoid backlash) and the combined influences of transmitted force and rotational velocity.

---

3 These values depend on the type of material and lubricant.

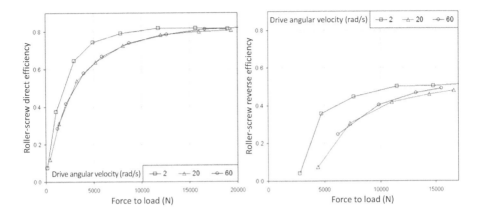

Figure 6.20. *Overall efficiency of a roller screw EMA (diameter 50 mm, lead 3 mm), according to [KAR 09]*

A lubrication fault, overheating or a damaged contact surface can lead to rapid deterioration of efficiency and can cause transmission jamming. It is very common to associate several actuation paths in force summing, either in active/active mode or in active/passive mode, in order to actuate a single load. The jamming of one of the actuators leads to load blocking, which is generally not acceptable (for example, for most of the flight controls or landing gear steering). Despite the diversity of developed solutions (see section 6.1), jamming is a feared event, which is still today one of the major obstacles to the massive introduction of EMAs in aerospace. The declutching upstream of the nut–screw, in the rotational domain and at low torque, poses few problems, as it can quite easily be achieved. The difficulty stems from the nut–screw, the last element of the power chain, which needs to be treated in order to avoid it becoming a node of unacceptable failure for the actuation function. Several concepts can contribute to solving this problem.

6.2.3.1. *Improving nut–screw reliability*

The solution would ideally consist of improving reliability of nut–screw systems, so that the overall reliability required for the actuator can be reached. For example, it is possible to use new recirculation principles in ball screws, one of the main failure causes, as proposed in [BAB 10].

6.2.3.2. *Anticipating jamming*

Health Monitoring is a promising path towards anticipating jamming. It has two stages. The first one is the diagnosis, which involves the earliest detection of the appearance of faults that become manifest in the actuator behavior. The second one is the prognosis, which predicts the future evolution of faults. Based on prognosis, immediate actions are undertaken (for example, reconfiguration or rerouting) or planned (for example, unscheduled maintenance) to prevent the fault from inducing severe failure. The "by-wire" electrical technology allows the low-cost implementation of Health Monitoring functions, as sensors, processing capacity and memory are already installed for other functions. The difficulty stems from the level of accuracy and robustness of diagnosis and prognosis algorithms, which shall early detect the appearance of faults and shall exploit ageing models in order to predict the future evolution of the fault. As illustrated in Figure 6.21, the weak experience feedback on the use of roller screws in EMAs in aerospace context has generated a substantial flow of work dedicated to better understanding the origin and evolution of faults [TOD 12a, TOD 12b]. While most of the programs cited in section 6.1 also approached health monitoring, numerous research works are still focusing on this subject, in order to improve readiness. Though a major subject, it is outside the scope of the present work.

Figure 6.21. *Jamming tests of a roller screw [TOD 12b]*

6.2.3.3. *Accepting the jamming of an actuator*

Instead of associating several actuators (or using a redundant actuator) for a single load, a further solution consists of splitting the load and associating each part of it with a single actuator (see Figure 6.11(b)). In the event of jamming, the other parts of the load are positioned to compensate the effect of the blocked part. In flight controls, splitting is already widely used for spoilers. It was implemented on a Boeing B787, as a first step in introducing EMAs for 4 of the 14 spoilers. The splitting solution may pose the problem of load position synchronization. It can also be seen as an opportunity to eliminate the force equalization problems or to provide new degrees of freedom allowing for performance improvement (for example, load reduction or improvement of aerodynamic efficiency).

6.2.3.4. *Declutching the load from the actuator in the event of jamming*

The nut–screw can be rendered jamming-tolerant if it can be disconnected from the actuated load. This approach is used in the redundant architectures with force summing for which the actuation function shall be realized when one of the paths is jammed (fail-functional or fail-operative actuation). It is also used in the architecture of simplex actuation when it is sufficient to release the load, as it is the case for the free fall of landing gear under gravity. The main difficulties relate to availability and testability, as well as to the declutching capacity in the presence of the load: in normal mode, the disconnection function, which is integrated in series on the force (or force reaction) path, shall transmit the full actuation forces.

a) *Mechanical fuse.* The mechanical fuse is the part joining two bodies that breaks when subjected to excessive force. It is an easy-to-implement solution currently used for the equipment (for example, hydraulic pump) installed on an engine's gearbox. It is of little interest for the actuation of critical functions, as it is not testable, is irreversible and requires maintenance work after fuse breaking.

b) *Controlled device.* There are two large families of controlled devices: irreversible or reversible.

The first family is based on pyrotechnic actuation. The separation of the two mechanical parts is electrically activated by firing a pyrotechnic charge.

Currently used for the separation of launcher stages, this solution has the same testability drawbacks as the mechanical fuses. It had nevertheless been evaluated in the HEMAS program (see section 6.1.4). The adopted solution [NAU 13] is presented in Figure 6.22. Under the effect of the gas pressure generated by the firing of the pyrotechnic device, the body (15) translates axially and frees the pins (3) that join the ball nut (1) in the nut–screw and the output shaft (4) linked to the driven load. With a mass of 1 kg, this device was functional for a nut transmitting a force of 40 kN.

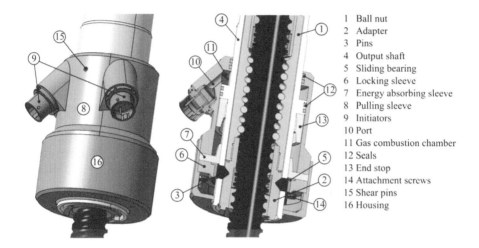

Figure 6.22. *Pyrotechnic declutch [NAU 13]*

The second family, which is electrically controlled, uses an electromechanical or electromagnetic actuator to release the load in case of nut–screw jamming. Depending on its design, this device can be testable and resettable (after elimination of the fault and setting into normal configuration). In the approach chosen for the landing gear extension/ retraction actuator in the CISACS project (see section 6.1.3.2), the primary gear assembly output in normal mode drives the screw into rotation using a key. The screw is fitted with a keyway throughout its length. In case of jamming, the secondary actuator releases the axial thrust bearing of the screw, which can then translate at the same time as the nut that blocks it (Figure 6.23). It is worth noting in passing that the nut of the screw also plays the role of a hydraulic piston. Combined with the flow limiter and the accumulator, it allows the realization of the damping function during free fall.

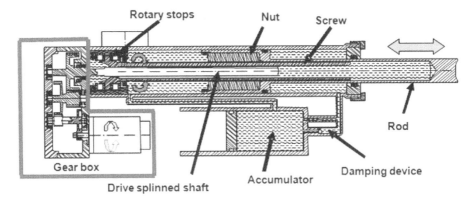

Figure 6.23. *Declutching by axial thrust bearing release, according to [MAR 11]*

A further solution consists of inserting, between the functionally translational element of the nut–screw and the load, a rod that can be separated from this element in the event of actuator failure. Many developments have been based on this principle, as illustrated in Figure 6.24.

6.2.3.5. *Rendering the nut–screw redundant*

The nut–screw can be rendered jamming-resistant if two nut–screws are associated in series on the power chain. As the nut of one of the systems is also the screw of the other, a telescopic cylinder is formed. In normal mode, only one of the nut–screw systems (normal path) operates: the screw and the nut of the second system (backup path) are joined through rotation. When the normal path is jammed, the backup path is activated by releasing the relative rotational motion of its screw and nut. In order to limit the length of the actuator, it is of interest to nest the two nut–screw systems, as illustrated in Figure 6.25 for extension/retraction EMA in the MELANY project (see section 6.1.3.2).

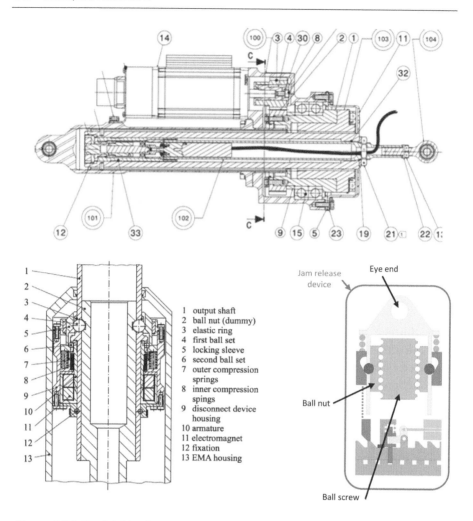

Figure 6.24. *Declutching by separation of the translational element of the nut–screw and the load, upper image: [JIM 12, JIM 16], middle image: [NAU 16], lower image: [BIL 14]. For a color version of this figure, see www.iste.co.uk/mare/aerospace2.zip*

6.2.4. *Breakage*

In some cases, such as the actuators of THS or even still the pylon conversion for tiltrotor crafts, the mechanical breaking of the force path between the actuator support structure and the driven load is a catastrophic event. In this case, the whole power path must be duplicated (particularly in terms of nut–screw) and the actuator must be rendered irreversible.

Figure 6.26 presents the mechanical architecture of a THS actuator. Two force paths (normal and backup) can be identified. The "no-back" in charge with the irreversibility function authorizes the power transfer from EMA to the load when it opposes the movement. If the load is aiding, the no-back realizes a braking function, blocking it relative to the actuator body. In order to move an aiding load, the actuator motor(s) must also inject power in the no-back.

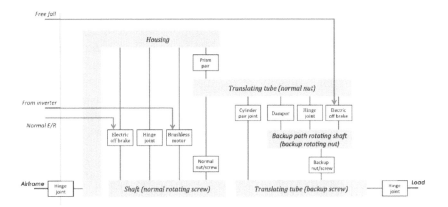

Figure 6.25. *Jam-tolerant nut–screw, according to [CHE 10]. For a color version of this figure, see www.iste.co.uk/mare/aerospace2.zip*

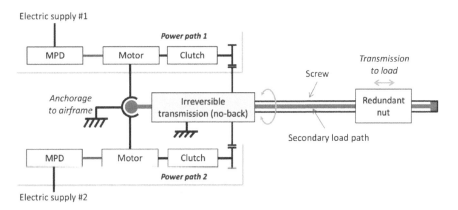

Figure 6.26. *Breaking-resistant EMA. For a color version of this figure, see www.iste.co.uk/mare/aerospace2.zip*

6.2.5. *Thermal equilibrium*

Similar to EHAs, the thermal equilibrium of EMAs determines their reliability and service life. Modeling the mechanical losses and exchanges between various bodies is similarly difficult. Numerical simulation software can certainly help when taking into account increasingly higher levels of detail of the many internal and external coupling effects [PAS 14, LEN 16]. Simulation results are unfortunately still too sensitive to uncertainties related to the model parameters (contact resistance, exchange coefficient when lubricant is partially present, etc.) to be able to provide decision elements, rather than comparison elements. The EMAs cannot use the fluid flowing through EHAs to improve the exchange with the environment. On the other hand, the high thermal capacity of the metallic components of the actuator allows the storage of heat during transients, as illustrated by the thermography image in Figure 6.27 captured near the rod of an EMA, 30 min after a full load sinus test (temperatures vary from 20°C for dark blue to 120°C for bright red).

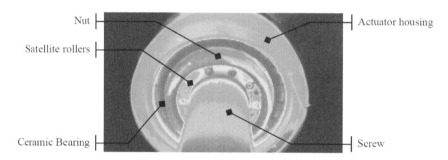

Figure 6.27. *Thermography image of an EMA [GRA 04]. For a color version of this figure, see www.iste.co.uk/mare/aerospace2.zip*

6.2.6. *Control*

Similar to EHAs, the generic architecture of EMA position control involves three nested loops of current, speed and position. On the other hand, it has certain specificities.

6.2.6.1. *Sequences and operating modes*

For the end-stop to end-stop actuation functions, control involves several sequences corresponding to various operation modes. For example, for an extension function, the sequence can involve:

– a current control to generate a retraction force that relieves the lock, to release it more easily;

– an acceleration phase up to the rated speed of extension, followed by a constant speed phase. The acceleration phase is initiated after time out, once the current reaches a predefined value;

– a deceleration phase, which is initiated by a position sensor, indicating the approach of mechanical stop;

– after arrival in end-stop at low speed, a phase of current increase in order to produce an extension force that facilitates engaging the lock.

6.2.6.2. *Dynamics of external forces rejection*

The position sensor generally used for control measures the rod position relative to the actuator body. The objective is, however, to position the driven load relative to the actuator support structure. As they are designed under mass constraint, the bodies are subjected to non-negligible deformation (several mm under rated force) at the level of the actuator body anchorage to the support structure and of the actuator rod transmission to the driven load. Load inertia, combined with these parasitic compliances, produces a parasitic dynamics (load mode) that is not sensed by the position sensor integrated in the actuator. Furthermore, as already mentioned, the measured current, which could ideally offer an image of the instantaneous force developed at the EMA/load interface, is altered by friction and inertia effects. The control performance of this mode is therefore low, as it is for realizing overload protection. While load mode varies little, it is possible to fit in a rejecting filter in series on the direct chain of control, similar to the arrangement for the conventional flight controls of certain commercial aircraft. A much more efficient solution that can also be used for overload protection consists of adding a force loop [SHI 01, DÉE 07]. It requires the installation of a sensor force at the EMA/load interface (similar to COVADIS demonstrator, see section 6.1.2.1). When the load ambient is too harsh, an alternative consists of inserting a sensor at the level of the EMA anchorage to the actuator support structure, as on the EMAs for the VEGA launcher (see section 6.1.1)[4]. The force loop is implemented as shown in Figure 6.28. In the form of pure proportional feedback, it unfortunately increases the static position error under permanent load. The force feedback is therefore affected by high-pass filtering (DFF for Dynamic Force Feedback) that eliminates this secondary effect. In a redundant architecture that uses two actuators simultaneously active on a single load, the force sensors can also be used for a force equalization function. This function is particularly significant if the two actuators have different static and dynamic behaviors, for example, because they use dissimilar technologies [WAN 14].

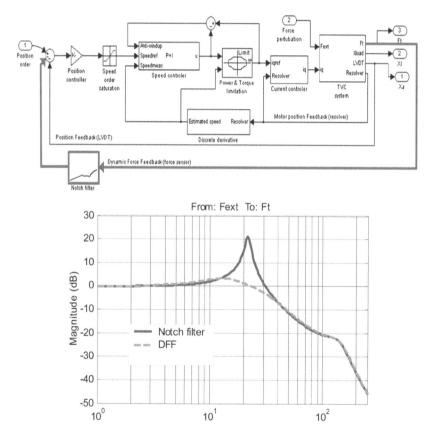

Figure 6.28. *Dynamic force feedback, from [DÉE 07]. For a color version of this figure, see www.iste.co.uk/mare/aerospace2.zip*

---

[4] But the force information is different in this case, as it involves the inertia force of the EMA housing.

6.2.7. *Further considerations*

There are still many considerations and important EMA evolution paths to be explored in terms of architectures, technological principles or even materials. Besides the advances in the fields of power electronics and electric machines, EMAs still have great potential for evolution at all levels: integration in the actuator support structure, distributed actuation, rotational actuators [RIV 16], dry lubrication, magnetic reducers [BEN 16], etc. For primary flight controls, it is estimated that the cost and mass of an EMA are still twice those of a conventional SHA actuator (but they typically comprise half the number of parts in an EHA). This gap should be rapidly bridged thanks to the efforts of standardization, modularization and mutualization, combined with the increased volume of manufactured EMAs.

Bibliography

[ACE 92] ACEE H., The integrated actuator package approach to primary flight control, SAE Technical Paper 920968, SAE Aerospace Atlantic, p. 10, 7–10 April 1992.

[AIM 10a] AIM GmbH, "ARINC 429 Specification Tutorial", vol. 2.1, p. 21, available at: https://www.aim-online.com/pdf/OVIEW429.PDF, November 2010.

[AIM 10b] AIM GmbH, "MIL-STD-1553 tutorial", vol. 2.3, p. 82, available at: https://www.aim-online.com/pdf/OVW1553.PDF, November 2010.

[AIR 13] AIRBUS, "Special Edition A350 XWB", FAST, Airbus Technical Magazine, available at: http://www.airbus.com/support/publications/?eID=maglisting_push&tx_maglisting_pi1%5BdocID%5D=41228, June 2013.

[AKI 16] AKIKI P., HAGE-HASSAN M., VANNIER J.-P. *et al.*, "Réduction des ondulations de couple d'un moteur à aimants en multi-V et bobinage sur dents", *Symposium de Génie Electrique (SGE 2016)*, Grenoble, June 2016.

[ALD 93] ALDEN R., "Flight demonstration, evaluation and proposed applications for various all electric flight control actuation systems concepts", *AIAA Aerospace Design Conference*, Irvine, CA, p. 11, 16–19 February 1993.

[ALE 10] ALEXANDRE P., TELTEU D., BAUDART F. *et al.*, "Failure operation of a multiphase drive for an electro mechanical actuator", *Proceedings of the 4th International Conference Recent Advances in Aerospace Actuation Systems and Components*, Toulouse, France, pp. 111–115, 5–7 May 2010.

[ARG 08] ARGILE R.N., MEEROW B.C., ATKINSON D.J. *et al.*, "Reliability analysis of fault tolerant drive topologies", *4th IET Conference on Power Electronics, Machines and Drives*, New York, pp. 1–15, 2–4 April 2008.

[ARN 98] ARNAUD A., "An approach to EHA standardization", *Proceedings of the International Conference on Recent Advances in Aerospace Hydraulics*, Toulouse, France, pp. 105–115, 24–26 November 1998.

[BAB 10] BABINSKI J., "Jam tolerant electro-mechanical actuators for aircraft flight and utility control", *Proceedings of the 4th International Conference on Recent Advances in Aerospace Actuation Systems and Components*, Toulouse, France, pp. 42–46, 5–7 May 2010.

[BAI 07] BAI H., "Analysis of a SAE AS5643 Mil-1394b based high-speed avionics network architecture for space and defense applications", *IEEE Aerospace Conference*, no. 1344, pp. 1–9, 3–10 March 2007.

[BAT 12] BATES L.B., YOUNG D.T., "Developmental testing of electric thrust vector control systems for manned launch vehicle applications", *Proceedings of the 41st Aerospace Mechanisms Symposium, Jet Propulsion Laboratory*, 6–18 May 2012.

[BEN 10] BENNETT J.W., Fault tolerant electromechanical actuators for aircraft, PhD Dissertation, Newcastle University, November 2010.

[BEN 16] BENAROUS M., TREZIÈRES M., "Magnetic gear-box for aerospace application", *Proceedings of the 7th International Conference on Recent Advances in Aerospace Actuation Systems and Components*, Toulouse, France, pp. 141–144, 16–17 March 2016.

[BET 15] BETTINI P., VAN DEN BOSSCHE D., "Weight reduction in cockpit controls: contributions since the A320", *SAE A-6 Spring Meeting*, Charleston, 10–14 May 2015.

[BIC 03] BICKEL N., BUTTER U., HAMMERLINDL M. *et al.*, "Getting a primary Fly-by-Light control system into flight", *American Helicopter Society 59th Annual Forum*, Phoenix, 6–8 May 2003.

[BIE 04] BIEDERMANN O., BILDSTEIN M., "Development, qualification and verification of the A380 spoiler EBHA", *Proceedings of the 2nd International Conference on Recent Advances in Aerospace Actuation Systems and Components*, Toulouse, France, pp. 97–101, 24–26 November 2004.

[BIL 98] BILDSTEIN M., "EHA for flight testing on Airbus A321 – power losses of fix pump EHA versus variable pump EHA", *Proceedings of the International Conference on Recent Advances in Aerospace Hydraulics*, Toulouse, France, pp. 101–103, 24–26 November 1998.

[BIL 14] BILDSTEIN M., "A built-in jam release device for electromechanical actuators in flight control", *Proceedings of the 6th International Conference on Recent Advances in Aerospace Actuation Systems and Components*, Toulouse, France, pp. 105–108, 2–3 April 2014.

[BOU 11] BOUYSSOU S., "Optical fibre on aircraft – when the light speed serves data transmission", *FAST Flight Airworthiness Support Technology Airbus Technical Magazine*, pp. 14–20, January 2011.

[BOW 04] BOWER J., "A380 Hydraulic Slat Drive Channel", *Proceedings of the 2nd International Conference on Recent Advances in Aerospace Actuation Systems and Components*, Toulouse, France, pp. 91–96, 24–26 November 2004.

[BUT 07] BUTZ H., "The Airbus approach to open integrated modular avionics (IMA) technology: technology, methods, processes and future road map", *Proceedings of the 1st Workshop on Aircraft Systems Technology (AST 07)*, Hamburg, Germany, pp. 211–221, 30 March 2007.

[CAO 12] CAO W., MECROW B.C., ATKINSON G.J. *et al.*, "Overview of electric motor technologies used for more electric aircraft (MEA)", *IEEE Transactions of Industrial Electronics*, vol. 59, no. 9, pp. 3523–3531, 2012.

[CAR 07] CARNEVALE C., RESTA P.D., "Vega electromechanical thrust vector control development", *43rd AIAA/ASME/ASEE Joint Propulsion Conference and Exhibit*, Cincinatti, p. 10, 8–11 July 2007.

[CHE 10] CHEVALIER P.-Y., GRAC S., LIEGEOIS P.-Y., "More electrical landing gear actuation systems", *Proceedings of the 4th International Conference on Recent Advances in Aerospace Actuation Systems and Components*, Toulouse, France, pp. 9–16, 5–7 May 2010.

[CHI 14] CHICO P., "Electric brake", *Proceedings of the 6th International Conference on Recent Advances in Aerospace Actuation Systems and Components*, Toulouse, France, pp. 25–28, 2–3 April 2014.

[COL 99] COLLINS A., "EABSYS Electrically Actuated Braking System", *Colloquium on Electrical Machines and Systems for the More Electric Aircraft*, pp. 4.1–4.5, 9 November 1999.

[COL 11] COLLINSON R.P.G., *Introduction to Avionics Systems*, 3rd ed., Springer, 2011.

[CON 04] CONARD J.-P., GILSON E., LABRIQUE F., "Sensorless speed control of an asynchronous motor – a military aircraft application", *Proceedings of the 2nd International Conference on Recent Advances in Aerospace Actuation Systems and Components*, Toulouse, France, pp. 119–127, 24–26 November 2004.

[COU 16] COUGO B., CARAYON J.-P., DOS SANTOS V. *et al.*, "Impacts of the use of SiC semiconductors in actuation systems", *Proceedings of the 7th International Conference on Recent Advances in Aerospace Actuation Systems and Components*, Toulouse, pp. 201–206, 16–18 March 2016.

[COW 93] COWAN J.R., WAIR R.A., "Design and test of electromechanical actuators for thrust vector control", *27th Aerospace Mechanisms Symposium*, pp. 349–366, 7 May 1993.

[CRA 01] CRANE/HYDRO-AIRE, "Autobrake", Brochure Commerciale, p. 2, 2001.

[CRA 08] CRANE D., *Aviation Maintenance Technicians Series: Airframe*, 3rd ed., Aviation Supplies and Academics Inc., 2008.

[CYC 81] CYCON M.F., "A dual input actuator for fluidic backup flight control", *WA-8B AICE Joint Automatic Control Conference*, Charlottesville, VA, vol. 1, no. 7, pp. 17–19 June 1981.

[DAM 08] D'AMORE M., GIGLIOTTI K., RICCI M. *et al.*, "Feasibility of broadband power line communication aboard an aircraft", *International Symposium on Electromagnetic Compatibility*, Hamburg, Germany, pp. 1–6, 8–12 September 2008.

[DAN 07] DANIEL J.-P., Fly-by-Wireless – Airbus End-user Viewpoint, CANEUS/NASA Workshop, Grapevine, TX, pp. 26–28, March 2007.

[DAN 15] DANIEL G., TAGLIANVINI A., MOSCA P., *Ouvrage 21-A: Connaissance Générale des aéronefs – tome 1*, Editions Jean Mermoz, Paris, 2015.

[DAU 14] DAUPHIN-TANGUY G., MARÉ J.-C., DENIS R. *et al.*, "Approche virtuelle pour la conception et le développement de systèmes mécatroniques – Méthodologie", *Techniques de l'Ingénieur, série Ingénierie des systèmes et robotique*, S-7800, pp. 23, 10 June 2014.

[DAV 15] DAVIDON W., ROIZES J., "Electro-hydrostatic Actuation System for aircraft landing gear actuation", *Proceedings of the 5th International Workshop on Aircraft Systems Technologies*, Hamburg, Germany, pp. 3–12, 24–25 February 2015.

[DEL 10] DE LA CHEVASNERIE A., GRAND S., LEGRAND B. *et al.*, "Electromechanical actuator/MOET project", *Proceedings of the 4th International Conference on Recent Advances in Aerospace Actuation Systems and Components*, Toulouse, France, pp. 83–87, 5–7 May 2010.

[DED 11] DE DONCKER R., PULLE D.W.J., VELTMAN A., *Advanced Electrical Drives – Analysis, Modeling, Control*, Springer, Netherlands, 2011.

[DÉE 07] DÉE G., VANTHUYNE T., ALEXANDRE P., "An electrical thrust vector control system with dynamic force feedback", *Proceedings of the 3rd International Conference on Recent Advances in Aerospace Actuation Systems and Components*, Toulouse, France, pp. 75–79, 13–15 June 2007.

[DÉG 10] DÉGARDIN V., SIMON E.P., MORELLE M. *et al.*, "On the possibility of using PLC in aircraft", *IEEE International Symposium on Power Line Communications and Its Applications*, Rio de Janeiro, Brazil, pp. 337–340, 28–31 March 2010.

[DEL 04] DELLAC S., TERNISIEN D., "Airbus 380 Electro-Hydraulic back-up architecture for Braking and Steering Systems", *Proceedings of the 2nd International Conference on Recent Advances in Aerospace Actuation Systems and Components*, Toulouse, France, pp. 103–108, 24–26 November 2004.

[DER 11] DERRIEN J.-C., TIEYS P., SENEGAS D. *et al.*, "EMA Aileron COVADIS Development", SAE AeroTech, Toulouse, France, p. 12, 21 October 2011.

[DES 12] DESCAMPS D., ALEXANDRE P., TELTEU-NEDELCU T., "Hi-rel electromechanical thrust vector actuation systems for European unmanned launch vehicles – a challenge for the next generation", *Proceedings of the 5th International Conference on Recent Advances in Aerospace Actuation Systems and Components*, Toulouse, France, pp. 11–158, 13–15 June 2012.

[DOG 11] DOGAN H., WURTZ F., FOGGIA A. *et al.*, "Analysis of Slot-Pole Combination of Fractional-slots PMSM for Embedded Applications", *ACEMP 2011*, Turkey, pp. 627–631, September 2011.

[DÖN 10] DÖNNEMZER Y., ERGENE L.T., "Skewing effect on interior type BLDC motors", *19th International Conference on Electrical Machines*, Rome, Italy, p. 5, 6 September 2010.

[DOR 07] DORKEL A., BIEDERMANN O., WENNMACHER G., "Theoretical validation and experimental verification of the electrically assisted hydraulic actuator concept", *Proceedings of the 3rd International Conference on Recent Advances in Aerospace Actuation Systems and Components*, Toulouse, France, pp. 15–19, 13–15 June 2007.

[DOW 09] DOWNING C., "Modular Approach Eases Avionics Certification Challenges", available at: http://www.cotsjournalonline.com/articles/print_article/101451, 2009.

[DUB 13] DUBUIS D., ALAIN TAGLIAVINI A., MOSCA P., "Connaissance générale des aéronefs Tome 3, Electricité, Equipements de secours/021", Institut Jean Mermoz, available at: www.institut-mermoz.com/contents/Ouvrages-de-cours-26.html, April 2013.

[EDG 78] EDGE J.T., An electromechanical actuator technology development program, SAE Technical Paper 780581, 1978.

[ELR 10] EL REFAIS A.M., "Fractional-slot concentrated windings synchronous permanent magnet machines: opportunities and challenges", *IEEE Transaction on Industrial Electronics*, vol. 57, no. 1, pp. 107–212, January 2010.

[ELG 10] ELGEZABAL O., Fly-by-Wireless (FBWSS): Benefits, risks and technical challenges, Fly-by-Wireless Workshop, Orono, ME, 24 August 2010.

[ELL 16] ELLIOTT N., LINFORTH S., MOORE C., "Thermae II (Main landing gear & door EH actuation system) – integration and testing", *Proceedings of the 7th International Conference on Recent Advances in Aerospace Actuation Systems and Components*, Toulouse, France, pp. 112–117, 16–17 March 2016.

[ELM 03] ELMENREICH W., IPP R., Introduction to TTP/C and TTP/A, Workshop on Time-Triggered and Real-Time Communication Systems, Manno, Switzerland, December 2003.

[EMP 13] EMPRINGHAM L., KOLAR J.W., RODRIGUEZ J. *et al.*, "Technological issues and industrial application of matrix converters: a review", *IEEE Transactions on Industrial Electronics*, vol. 60, no. 10, pp. 4260–4271, October 2013.

[EUR 12] EUROPEAN AVIATION SAFETY AGENCY, "Certification Specifications for Large Rotorcraft – CS29", Amendment 3, available at: https://www.easa.europa.eu, 11 December 2012.

[EUR 15] EUROPEAN AVIATION SAFETY AGENCY, "Certification Specifications and Acceptable Means of Compliance for Large Aeroplanes – CS25", Amendment 17, 15 July 2015.

[FAU 93] FAULKNER A., "Smart actuator systems: a practical solution?", *Proceedings of the Aerospace Hydraulics and Systems International Conference (IMECH 1993 6)*, London, pp. 123–132, 29–30 September 1993.

[FOC 11] FOCH H., LADOUX P., PIQUET H., "Association de cellules de commutation – Éléments de synthèse des convertisseurs statiques", *Encyclopédie des techniques de l'ingénieur*, 10 May 2011.

[FUE 07] FUERST D., NEUHEUSER T., "Development, prototype production and testing of an electromechanical actuator for a swashplateless primary and individual helicopter balde control system", *Proceedings of the 1st Workshop on Aircraft Systems Technology*, Hamburg, Germany, pp. 7–19, 29–30 March 2007.

[FUL 96] FULMER C., 40 HP Electro-mechanical actuator, NASA Contractor Report 198509, October 1996.

[GIA 14] GIANFRANCESCHI M., JACAZIO G., WANG J., "High bandwidth electromechanical actuator for swashplateless blade control system", *Proceedings of the 6th International Conference on Recent Advances in Aerospace Actuation Systems and Components*, Toulouse, France, pp. 1–8, 2–3 April 2014.

[GIE 10] GIERAS J.F, *Permanent Magnet Motor Technology: Design and Applications*, 3rd ed., CRC Press, 2010.

[GOD 02] GODO E.L., "Flight control system with remote electronics", *Proceedings of the 21st Digital Avionics Systems Conference*, vol. 2, pp. 13B1-1–13B1-7, 2002.

[GRA 04] GRAND S., VALEMBOID J.-M., "Electromechanical actuators design for thrust vector control", *Proceedings of the 2nd International Conference on Recent Advances in Aerospace Actuation Systems and Components*, Toulouse, France, pp. 21–27, 24–26 November 2004.

[GRA 09] GRAHAM-ROWE D., "Fly-by-wireless set for take off", *New Scientist*, vol. 203, no. 2724, pp. 20–21, 2 September 2009.

[GRA 16] GRAND S., BALDUCCI G., FERVEL M. *et al.*, "Actuator control: a successful modular multi-application approach or Actuation2015 and beyond", *Proceedings of the 7th International Conference on Recent Advances in Aerospace Actuation Systems and Components*, Toulouse, France, pp. 60–65, 16–18 March 2016.

[GRE 88] GREENE J.B. *et al.*, Low energy consumption hydraulic techniques, Final Technical Report AD-A206 285, p. 298, 30 August 1988.

[GRE 97] GRELLET G., CLERC G., *Les actionneurs électriques. Principes, modèles, commande*, Eyrolles, 1997.

[GRE 04] GREISSNER C., CARL U., "Control of an electro-hydrostatic actuation system for the nose landing gear of an all electric aircraft", *Proceedings of the 2nd International Conference on Recent Advances in Aerospace Actuation Systems and Components*, Toulouse, France, pp. 9–13, 24–26 November 2004.

[GWA 06] GWALTNEY D.A., BRISCOE J.M., Comparison of communication architectures for spacecraft modular avionics systems, Report, NASA/TM-2006-214431, p. 36, June 2006.

[HAG 04] HAGEN J., MOORE L., ESTES J. *et al.*, "The X-38 V-201 flap actuator mechanism", *Proceedings of the 37th Aerospace Mechanisms Symposium, Johnson Space Center*, pp. 377–390, 19–21 May 2004.

[HAL 95a] HALSKI D., "Fly-by-light flight control systems", *Proceedings SPIE 2467, Fly-by-Light: Technology Transfer*, pp. 34–45, 22 May 1995.

[HAL 95b] HALSKI D., KESSLER B.L., MATTES R.E. *et al.*, "The fly-by-light aircraft closed loop flight test program", *Proceedings SPIE 2467, Fly-by-Light: Technology Transfer*, pp. 232–245, 22 May 1995.

[HAM 03] HAMMET R., "Flight-critical distributed systems: design considerations", *IEEE AESS Magazine*, pp. 30–36, June 2003.

[HAR 83] HARCHBURGER H.E., Development of redundant flight control actuation systems for the F/A-18 Strike Fighter, SAE Paper no. 831484, October 1983.

[HAR 95] HARRINGTON C., "High performance input servovalve", *SPIE 2467, Fly-by-Light: Technology Transfer*, pp. 195–204, 22 May 1995.

[HOL 96] HOLTZ J., SRPINGOB L., "Identification and compensation of torque ripple in high-precision permanent magnet motor drives", *IEEE Transactions on Industrial Electronics*, vol. 43, no. 2, pp. 309–320, April 1996.

[HOO 71] HOOKER D.S., KISSLINGER R.L., SMITH G.R. *et al.*, Survivable flight control system interim report no.1 – Studies, analyses and approach, Report AFFDL-TR-71-20, p. 375, May 1971.

[HSU 95] HSU J.S, SCOGGIN B.P., SCUDIERE M.B. *et al.*, Nature and Measurements of Torque Ripple of Permanent-magnet Adjustable-speed Motors, Oak Ridge National Laboratory, 1995.

[INS 02] INSTITUTE of ELECTRICAL AND ELECTRONICS ENGINEERS, "Standard for high performance serial bus (high speed supplement)", *IEEE Standard*, 2002.

[IOR 10a] IORDANIDIS G., DELLAC S., "Distributed and redundant electro mechanical nose gear steering system", *SAE International Journal of Aerospace*, vol. 2, no. 1, pp. 46–53, 2010.

[IOR 10b] IORDANIDIS G., "An overview of modelling and simulation activities for an all-electric nose wheel steering system", *Proceedings of the 4th International Conference Recent Advances in Aerospace Actuation Systems and Components*, Toulouse, France, pp. 145–153, 5–7 May 2010.

[JIA 14] JIAN Z., YANGWEI Y., "Brushless DC motor fundamentals application note", AN047 Rev. 1.0, available at: www.monolithicpower.com, 7 May 2014.

[JIM 12] JIMENEZ A., NOVILO E., AGUADO E. *et al.*, "Electromechanical actuator with anti-jamming system for safety critical aircraft applications", *Proceedings of the 5th International Conference on Recent Advances in Aerospace Actuation Systems and Components*, Toulouse, France, pp. 77–83, 13–14 June 2012.

[JIM 16] JIMENEZ A., NOVILLO E., AGUADO F. *et al.*, "Electromechanical actuator with anti-jamming system for safety critical applications", *Proceedings of the 7th International Conference on Recent Advances in Aerospace Actuation Systems and Components*, Toulouse, France, pp. 27–32, 16–18 March 2016.

[JOM 09] JOMIER T., "Moet technical report", available at: www.transport-research.info/.../20121218_094726_85827_MOET_Public_Technical, 14 December 2009.

[JON 06] JONES C.H., "Communications over aircraft power lines", *IEEE International Symposium on Power Line Communications and Its Applications*, Orlando, FL, pp. 149–154, 2006.

[JOR 10] JORGE Z., DEB D., "Sensorless field oriented control of a PMSM", Microship application note 1078, p. 28, available at: www.microchip.com/downloads/en/AppNotes/01078B.pdf, 2010.

[KAR 00] KARNOPP D.C., MARGOLIS D.L., ROSENBERG R.C., *System Dynamics – Modelling and Simulation of Mechatronic Systems*, 3rd ed., John Wiley & Sons, New York, 2000.

[KAR 06] KARAM W., MARÉ J.-C., "Comparison of EMA et HA performance for dynamic load simulators", *Proceedings of Bath Power Transmission and Motion Control Symposium*, Bath, England, pp. 211–224, 7–9 September 2006.

[KAR 09] KARAM W., MARÉ J.-C., "Modelling and simulation of mechanical transmission in roller-screw electromechanical actuators", *Aerospace Engineering and Aircraft Technology Journal*, vol. 81, no. 4, pp. 288–298, 2009.

[KOH 80] KOHNHORST L.L., MAGNACCA D.A., Design and test of a hydra-optic flight control actuation system (HOFCAS) concept, NAVAIR DEVCEIM-79156-60 Report, December 1980.

[KOL 08] KOLIATENE F., LEBEY T., CAMBRONNE J.P. et al., "Impact of the aeronautic environment on the partial discharges ignition: a basic study", *Conference Record of the 2008 IEEE International Symposium on Electrical Insulation*, ISEI, 2008.

[KOP 01] KOPALA D.J., DOELL C., "High performance electromechanical actuation for primary flight surfaces (EPAD program results)", *Proceedings of the 1st International Conference on Recent Advances in Aerospace Actuation Systems and Components*, Toulouse, France, pp. 71–76, 13–15 June 2001.

[KOP 04] KOPETZ H., "From a federated to an integrated architecture for dependable embedded systems", *Proceedings of the 8th Annual High Performance Embedded Computing (HPEC) Workshops*, vol. 1, 28–30 September 2004.

[KRI 10] KRISHNAN R., *Permanent Magnet Synchronous and Brushless DC Motor Drives*, 1st ed., CRC Press, 2010.

[KUL 07] KULSHRESHTHA A., "Remote actuation control system: aircraft flight control for hydraulic-servo and electric actuation", *Proceedings of the 3rd International Conference on Recent Advances in Aerospace Actuation Systems and Components*, Toulouse, France, pp. 155–162, 13–15 June 2007.

[LAC 99] LACROUX G., *Les actionneurs électriques pour la robotique et les asservissements*, Technique and Documentation, 1999.

[LAW 91] LAWHEAD P., Electro-modulated control of supply pressure in hydraulic systems, SAE Technical Paper 912119, p. 5, 1991.

[LEC 14] LECLEIR M., BARREMAECKER L., "Two phase cooling system for aircraft actuators", *Proceedings of the 6th International Conference on Recent Advances in Aerospace Actuation Systems and Components*, Toulouse, France, pp. 160–166, 2–3 April 2014.

[LEN 16] LENOBLE G., OLIVIR M., DNJAT D. et al., "Elevator actuator housing bay flight mission thermal integration analysis", *Proceedings of the 7th International Conference on Recent Advances in Aerospace Actuation Systems and Components*, Toulouse, France, pp. 118–126, 16–18 March 2016.

[LIE 12] LIEGEOIS P.-Y., NEVORET P., "Presentation of EMA for steering application", *Proceedings of the 5th International Conference on Recent Advances in Aerospace Actuation Systems and Components*, Toulouse, France, pp. 39–43, 13–14 June 2012.

[LOR 70] LORENZETTI R.C., HOOKER D.S., Survivable flight control system development program, Report FDC/ADPO-TM-70-1, p. 21, January 1970.

[MAR 99] MARÉ J.-C., MOULAIRE P., "Expert rules for the design of position control of electrohydraulic actuators", *Proceedings of the 6th Scandinavian International Conference on Fluid Power*, Tampere, pp. 1267–1280, 26–28 May 1999.

[MAR 04] MARÉ J.-C., "Les innovations de l'A380", *Revue Fluides, Numéro Spécial Matériels*, pp. 54–59, September 2004.

[MAR 06] MARÉ J.-C., "Contribution to the modelling, the simulation and the control synthesis of an aerospace electro-hydrostatic actuator", *Proceedings of the 5th International Fluid Power Conference*, Aachen, Germany, pp. 201–212, 20–22 March 2006.

[MAR 09] MARÉ J.-C., BUDINGER M., "Comparative analysis of energy losses in servo-hydraulic, electro-hydrostatic and electro-mechanical actuators", *The 11th Scandinavian International Conference on Fluid Power (SICFP'09)*, Linköping, Sweden, pp. 66–67, 2–4 June 2009.

[MAR 11] MARÉ J.-C., "Combining hydraulics and electrics for innovation and performance improvement in aerospace actuation", *Proceedings of the 12th Scandinavian International Conference on Fluid Power*, Tampere, Finland, 18–20 May 2011.

[MC 74] MC RUER D.T. et al., Mathematical models of human pilot behavior, National Technical Information Service, U.S. Department of Commerce, Paris, January 1974.

[MC 75] MC MAHON W., Apollo experience report – guidance and control systems: CSM service propulsion system gimbal actuators, NASA Technical Note TN D-7969, p. 14, July 1975.

[MEA 13] MEASUREMENT SPECIALTIES INC., The LVDT: construction and principles of operation, Technical paper, April 2013.

[MEI 08] MEIER F., Permanent magnet synchronous machines with non overlapping concentrated windings for low speed direct-drive applications, PhD Dissertation, Royan Institute of Technology, Stockholm, 2008.

[MIH 11] MIHAILA V., Nouvelle conception des bobinages statoriques des machines à courant alternatif pour réduire les effets négatifs des dV/dt, PhD Thesis, University of Artois, 14 December 2011.

[MIL 80] MILITARY SPECIFICATION, Flying qualities of piloted aiplanes MIL-F-8785C, 5 November 1980.

[MIL 97] MILITARY SPECIFICATION, Flying qualities of piloted aircrafts MIL-HDBK-1797, 19 December 1997.

[MOI 08] MOIR I., SEABRIDGE A., *Aircraft Systems: Mechanical, Electrical, and Avionics Subsystems Integration*, 3rd ed., Wiley, New York, 2008.

[MOI 13] MOIR I., SEABRIDGE A., JUKES M., *Civil Avionics Systems*, 2nd ed., Wiley, 2013.

[MON 96] MONTERO YANEZ J.F., *Advances in Onboard System Technology: All Electric Flight Control Actuation (ELAC)*, Wiley, 1996.

[MOO 01] MOORHOUSE D., MAXWELL C., BILDSTEIN M., "Electro hydrostatic actuator for primary flight control of very large aircraft", *Proceedings of the 1st International Conference on Recent Advances in Aerospace Actuation Systems and Components*, Toulouse, France, pp. 101–103, 55–58, June 2001.

[MOR 16] MORRA J., "Fiber optics transmit data and power over same cable", *Electronic Design*, 29 January 2016.

[MUL 10] MULTON B., "Les machines synchrones autopilotées", Préparation à l'agrégation de Génie Electrique, France, p. 62, 2010.

[NAU 13] NAUBERT A., CRHISTMANN M., JANKER P. *et al.*, "Anti-jamming mechanism for electromechanical actuators", *Proceedings of the 4th International Workshop on Aircraft System Technologies*, Hamburg, Germany, pp. 139–147, 23–24 April 2013.

[NAU 16] NAUBERT A., BINZ H., BACHMANN M. *et al.*, "Disconnect device design options for jam-tolerant electromechanical actuators", *Proceedings of the 7th International Conference on Recent Advances in Aerospace Actuation Systems and Components*, Toulouse, France, pp. 187–192, 16–18 March 2016.

[NAV 97] NAVARRO R., Performance of an electro-hydrostatic actuator for the F-16 systems research aircraft, Report, NASA-TM-97-206224, p. 38, October 1997.

[NOR 86] NORTON W.J., Advanced electromechanical actuation system, Report AD-A176 148, Air Force Wright Aeronautical Laboratories, June 1986.

[NOV 99] NOVACEK G., "Accurate linear measurement using LVDTs", *Circuit Cellar Ink*, no. 106, pp. 20–27, May 1999.

[OBE 12] OBERMAISSER R., *Time Triggered Communication*, CRC Press, 2012.

[OBR 08] O'BRIEN J., KULSHRESHTHA A., "Distributed and remote control of flight control actuation using power line communications", *IEEE/AIAA 27th Digital Avionics Systems Conference*, Saint Paul, MN, 26–30 October 2008.

[PAS 14] PASIES-RUBERT A., MUR C., GARAY M. *et al.*, "Benefits of multiphysics integration through cosimulation – case study: heat monitoring on a primary flight control actuator", *Proceedings of the 6th International Conference on Recent Advances in Aerospace Actuation Systems and Components*, Toulouse, France, pp. 144–149, 2–3 April 2014.

[RAO 09] RAO T., "CAN Bus in Aviation", *Avionics Magazine*, May 1, 2009.

[RAS 11] RASHID M.H., *Power Electronics Handbook. Devices, Circuits and Applications*, 3rd ed., Elsevier, 2011.

[RAY 93] RAYMOND E.T., CHENOWETH C.C., *Aircraft Flight Control Actuation System Design*, SAE Press, Warrendale, 1993.

[RIV 16] RIVIERA S., ANDREOU A., "Rotary actuator for retractable landing gear systems", *Proceedings of the 7th International Conference on Recent Advances in Aerospace Actuation Systems and Components*, Toulouse, France, pp. 133–139, 16–18 March 2016.

[ROA 97] ROACH J.M., "FLASH electrohydrostatic actuation modelling, analysis and test results", *SAE Aerospace Power Systems Conference Proceedings, Paper 971234*, 1997.

[ROB 05] ROBOAM X., LANGLOIS O., MARÉ J.-C. *et al.*, "Bond Graph modelling of an electrohydrostatic actuator for aeronautic application", *Bond Graph Methodology – Theory and applications' of the IMAACS World Congress*, Paris, 15–17 July 2005.

[ROS 03] ROSKAM J., *Airplane Flight Dynamics and Automatic Flight Controls*, Darcorporation, 2003.

[ROT 14] ROTTACH M., GERADA C., WHEELER P., "Helicopter EMA system: electrical drive optimization and test", *Proceedings of the 6th International Conference on Recent Advances in Aerospace Actuation Systems and Components*, Toulouse, France, pp. 9–13, 2–3 April 2014.

[SAH 15] SAHA S., CHO Y.-H., "Design of 3-step skew BLAC motor for better performance in electric power", *International Journal of Mechanical Aerospace, Industrial Mechatronic and Manufacturing Engineering*, vol. 9, no. 7, pp. 1264–1269, 2015.

[SCH 93] SCHAEFFER W.S., INDERHEES L.J., MOYNES J.F., "Flight control actuation for the B-2 advanced technology bomber", *Proceedings of the Aerospace Hydraulics and Systems International Conference (IMECH 1993-6)*, London, pp. 23–32, 29–30 September 1993.

[SCH 98] SCHMITT R.S., MORRIS J.R., *Fly by Wire*, SAE Press, Warrendale, 1998.

[SCH 08] SCHUSTER T., VERMA D., "Networking concepts comparison for avionics architecture", *Proceedings of the 27th IEE Digital Avionics Systems Conference*, p. 11, 26–30 October 2008.

[SEE 12] SEEMANN S., CHRISTMANN M., JANKER P., "Control and monitoring concept for a fault-tolerant electromechanical actuation system", *Proceedings of the 5th International Conference on Recent Advances in Aerospace Actuation Systems and Components*, Toulouse, France, pp. 39–43, 13–14 June 2012.

[SHA 15] SHAH S.D., BUMATARIA R.K., CHOUDHARY A.M. *et al.*, "Primary flight control of Boeing 777", *International Journal of Advance Research in Engineering, Science & Technology*, vol. 2, no. 5, May 2015.

[SHE 85] SHERIF F.E., Advanced digital optical control actuation, SAE Technical Paper 851755, p. 13, 1 October 1985.

[SHI 01] SHINSTOCK D.E., SCOTT D.A., HASKEW T.A., "Transient force reduction in electromechanical actuator for thrust-vector control", *Journal of Propulsion and Power*, vol. 17, no. 1, pp. 65–72, January–February 2001.

[SKF 03] SKF, "The new SKF model for calculation of the frictional moment", *SKF General Catalog*, pp. 89–101, June 2003.

[SMI 96] SMITH R., Joint Strike Fighter integrated subsystems technology (J/IST) demonstration program overview, SAE Technical Paper 962259, 1996.

[SMI 07] SMITH T., Aircraft wiring reduction, Fly-by-Wireless Workshop, Grapevine, TX, 26–28 March 2007.

[SOC 04] SOCIETY OF MECHANICAL ENGINEERS, IEEE-1394b interface requirements for military and aerospace vehicle applications, Aerospace Standard SAE-AS5643, 30 December 2004.

[SOL 10] SOLTERO M., JING ZHANG J., COCKRIL C., RS-422 and RS-485 Standards overview and system configurations, Texas Instrument Application Report SLLA070D, May 2010.

[SON 16] SONCEBOZ, Compact power BLDC MD07210R1002, Technical documentation, available at: http://www.sonceboz.com/en/Compact-power-BLDC/, 2016.

[SPE 93] SPENCER E., "Development of variable pressure hydraulic systems for military aircraft utilizing the 'smart hydraulic pump'", *Proceedings of Aerospace Hydraulics and Systems IMechE Conference C474/002*, London, pp. 101–111, 20–30 September 1993.

[SPI 14] SPITZER C., FERRELL U., FERRELL T., *Digital Avionics Handbook*, 3rd ed., CRC Press, 2014.

[STI 04] STILES L.R., FREISNER A.L., MAYO J. *et al.*, "Impossible to resist: the development of rotorcraft Fly-by-Wire technology", *American Helicopter Society International, 60th Annual Forum*, Baltimore, MD, 7–10 June 2004.

[STO 89] STOCK M., KONING H., ZELLER S., "OPST 1 – an optical yaw control system for high performance helicopters", *American Helicopter Society Annual Forum*, Boston, 22–24 May 1989.

[TAK 04] TAKEBAYASHI W., HARA Y., "Thermal design tool for EHA", *Proceedings of the 2nd International Conference on Recent Advances in Aerospace Actuation Systems and Components*, Toulouse, France, pp. 15–19, November 24–26 2004.

[TAK 08] TAKAHASHI N., KONDO T., TAKADA M. *et al.*, "Developement of prototype electro-hydrostatic actuator for landing gear extension and retraction system", *Proceedings of the 7th JFPS International Symposium on Fluid Power*, Toyama, Japan, pp. 165–168, 2008.

[TER 89] TERRY J.L., DICKINSON J.D., JERAULD G.D., The advanced digital-optical control system (ADOCS) user demonstration program, US Army report USAAVSCOM TM-89-D-2, p. 45, September 1989.

[TES 93] TESKE D., FAULKNER D., "Electromechanical flight control servoactuator", *Intersociety Energy Conversion Engineering Conference*, Orlando, FL, pp. 1021–1025, 21–26 August 1993.

[TIA 10] TIAN H., TROJAK T., JONES C.H., "Communications over aircraft power lines: a practical implementation", *International Telemetering Conference Proceedings*, pp. 1–8, 2010.

[TOD 96] TODD J.R., "Fly-by-light flight control system development for transport aircraft", *15th Digital Avionics Systems Conference*, Atlanta, GA, pp. 153–158, 27–31 October 1996.

[TOD 98] TODD J.R., "A review of the fly-by-light optical aileron trim flight demonstration system", *Proceedings of the 17th Digital Avionics Systems Conference*, Bellevue, OH, 31 October–7 November 1998.

[TOD 07] TODESCHI M., "A380 flight control actuation – lessons learned on EHAs design", *Proceedings of the 3rd International Conference on Recent Advances in Aerospace Actuation Systems and Components*, Toulouse, France, pp. 21–26, 13–15 June 2007.

[TOD 12a] TODESCHI M., "Airbus – EMAs for flight controls actuation system – perspectives", *Proceedings of the 4th International Conference on Recent Advances in Aerospace Actuation Systems and Components*, Toulouse, France, pp. 1–8, 5–7 May 2012.

[TOD 12b] TODESCHI M., "Airbus – EMA for flight controls actuation system, 2012 status and perspectives", *Proceedings of the 5th International Conference on Recent Advances in Aerospace Actuation Systems and Components*, Toulouse, France, pp. 1–10, 13–14 June 2012.

[TOD 14a] TODESCHI M., "Airbus – Health Monitoring for the Flight Control EMAs: 2014 status and perspectives", *Proceedings of the 6th International Conference on Recent Advances in Aerospace Actuation Systems and Components*, Toulouse, pp. 73–83, 2–3, April 2014.

[TOD 14b] TODESCHI M., Power electronics for mechanical engineers, SAE-A6 Professional Development Seminar, Santa Barbara, 20–23 October 2014.

[TOD 16] TODESCHI M., SALAS F., "Power electronics for the flight control actuators", *Proceedings of the 7th International Conference on Recent Advances in Aerospace Actuation Systems and Components*, Toulouse, France, pp. 1–9, 16–18 March 2016.

[TRA 02] TRAINER D.R., WHITLEY C.R., "Electric actuation power quality management of aerospace flight control systems", *International Conference on Power Electronics, Machines and Drives*, pp. 229–234, 4–7 June 2002.

[TRA 06] TRAVERSE P., LACAZE I., SOUYRIS J., "Airbus Fly-by-Wire: a total approach to dependability", *Proceedings of the 25th International Congress of the Aeronautical Sciences*, Hamburg, Germany, p. 10, 3–8 September 2006.

[TRO 96] TROSEN D., CANNON B.J., "Electric actuation and control system", *Proceedings of the 31st Energy Conversion Engineering Conference*, Washington, DC, pp. 197–202, 11–16 August 1996.

[TUT 68] TUTT G.E., JANSEN J.A., "Dynamics of the Apollo electromechanical actuator", *Journal of Spacecraft*, vol. 5, no. 5, pp. 541–546, May 1968.

[USF 12] US FEDERAL AVIATION ADMINISTRATION, "Aviation Maintenance Technician Handbook – Airframe", FAA-H-8083-31, available at: https://www.faa.gov, 2012.

[VAL 14] VALDO M., Electro-hydrostatic actuators – a new approach in motion control, Workshop on Innovative Engineering for Fluid Power, Sao Paulo, Brazil, 2–3 September 2014.

[VAN 02] VAN DEN BOSSCHE D., "The evolution of the Airbus primary flight control actuation systems", *Proceedings of the 3rd Internationales Fluidtechnishes Kolloquium*, Aachen Germany, pp. 355–366, 5–6 March 2002.

[VAN 06] VAN DEN BOSSCHE D., "The A380 flight control electrohydrostatic actuators – achievement and lessons learnt", *25th Congress of International Council of the Aeronautical Sciences*, Paper ICAS 2006-7.4.1, Hamburg, Germany, p. 8, 3–8 September 2006.

[VAN 07] VAN BUITEN C., MC CABE B., The promise and challenge of wireless, Fly-by-Wireless workshop, Grapevine, TX, 26–28 March 2007.

[VAN 09] VANTHUYNE T., "An electrical thrust vector control system for the VEGA launcher", *Proceedings of the 13th European Space Mechanisms and Tribology Symposium*, Vienna, Austria, 23–25 September 2009.

[VAN 16] VAN DER LINDEN F., DREYER N., DORKEL A., "EMA health monitoring: an overview", *Proceedings of the 7th International Conference on Recent Advances in Aerospace Actuation Systems and Components*, Toulouse, France, pp. 21–26, 16–18 March 2016.

[VER 11] VERHOEVEN D., RENTÉ D., "Locking mechanism for IXV re-entry demonstrator flap control system", *14th European Space Mechanisms & Tribology Symposium*, Constance, Germany, pp. 329–336, 28–30 September 2011.

[VER 13] VERHOEVEN D., DE COSTER F., "Electro-mechanical actuators (EMA's) for space applications", *15th European Space Mechanisms & Tribology Symposium*, Noordwijk, The Netherlands, 25–27 September 2013.

[WAN 14] WANG L., MARÉ J.-C., "A force equalization controller for active/active redundant actuation system involving servo-hydraulic and electro-mechanical technologies", *Proceedings of the Institution of Mechanical Engineers, Part G: Journal of Aerospace Engineering*, vol. 228, no. 10, pp. 1768–1787, August 2014.

[WAR 15] WARWICK G., "Active sidestick controls make commercial debut – active inceptors gaining in popularity for commercial aircraft", *Aviation Week & Space Technology*, 1 February 2015.

[WEL 04] WELCHKO B.A., LIPO T.A., JAHNS T.M. *et al.*, "Fault tolerant three-phase AC motor drive topologies: a comparison of features, cost, and limitations", *IEEE Transaction on Power Electronics*, vol. 19, no. 4, pp. 1108–1116, July 2004.

[WHE 07] WHELLER P.W. *et al.*, "Design and reliability of a rudder EMA with an integrated permanent magnet machine and matrix converter drive", *Proceedings of the 3rd International Conference on Recent Advances in Aerospace Actuation Systems and Components*, Toulouse, France, pp. 21–26, 13–15 June 2007.

[WHI 07] WHITLEY C., ROPERT J., "Development, manufacture & flight test of spoiler EMA system", *Proceedings of the 3rd International Conference on Recent Advances in Aerospace Actuation Systems and Components*, Toulouse, France, pp. 215–220, 13–15 June 2007.

[WIL 08] WILD T.W., *Transport Category Aircraft Systems*, 3rd ed., Jeppesen, 2008.

[WIL 09] WILKINSON R., *Aircraft Structures and Systems*, 3rd ed., MechAero Publishing, St Albans, 2009.

[WRI 99] WRIGHT P., "Helicopter electro-mechanical actuation technology", *Colloquium on Electrical Machines and System for the More Electric Aircraft*, pp. 13.1–13.6, 9 November 1999.

[YEH 96] YEH Y.C., "Triple-triple redundant 777 primary flight computer", *Proceeding of IEE Aerospace Applications Conference*, Aspen, CO, vol. 1, pp. 293–307, 1996.

[YOU 45] YOUNG A.M., Aircraft U.S. Patent 2,384,516, available at: http://www.google.ch/patents/US2384516, 11 September 1945.

[ZAV 97] ZAVALA E., Fiber optic experience with the smart actuation system on the F-18 systems research aircraft, Report, NASA/TM-97-206223, p. 17, October 1997.

[ZIE 85] ZIEGLER B, DURANDEAU M., "Flight controls", *FAST Flight Airworthiness Support Technology – Airbus Technical Magazine*, Denis Dempster, no. 5, pp. 20–25, available at: www.airbus.com/support/publications/, May 1985.

Notations and Acronyms

Symbols

a, b, c	Phase axes	
B	Magnetic field	T
C	Capacitance	Fd
d	Disturbance	
E	Electromotive force	V
$\mathcal{E}$	Energy	J
f	Frequency	Hz
F	Force	N
I	(Line) Current	A
J	Phase current Moment of inertia	A m²kg
k	Factor	
K_m	Motor constant	Nm/A Vs/rad
l	Length	m
L	Inductance	H
m	Modulation ratio	
M	Mass	kg
N	Number of phases	

N	Number, Reduction ratio	
p	Number of pole pairs, Screw thread	m/tr
P	Pressure	Pa
$\mathcal{P}$	Power	W
q	Number of slots per pole	
Q	Volume flow rate	m³/s
r	Radius	m
R	Ratio, resistance	-, Ω
s	Slip ratio	
S	Area	m²
t	Time	s
T	Period, torque	s, Nm
u	Control	
U	(Phase) Voltage	V
v	Linear velocity	m/s
V	(Line) Voltage	V
V_0	Displacement	m³/rad
$w1$	Winding (first harmonic)	
x	Variable, position	-, m
y	Set point	
z	number of teeth	
α	Rotor frame axis, Modulation ratio	
β	Rotor frame axis	
δ	Electric angle	rad
Δ	Difference	
φ	Phase angle	rad
ψ	Torque control angle	rad

ϕ	*Magnetic flux*	Wb
λ	*Failure rate*	1/h
ρ	*Specific density*	kg/m^3
σ	*Electrical conductivity*	Ωm
τ	*Time constant*	s
ω	*Angular velocity, Angular frequency*	rad/s
θ	*Angular position*	rad

Exponents

*	*Setpoint*
^	*Mean*
γ	*Steinmetz exponent*

Indices

AC	*Alternative current*
a, b, c	*Actual indices*
c	*Cogging, conduction*
C	*Control*
com	*Switching*
con	*Conduction*
d	*Direct, diode*
D	*Disturbance*
DC	*Direct current*
e	*Input, electric, equivalent, excitation, electromagnetic*
E	*Eddy current*

f	*Filtered, friction*
h	*Hysteresis*
i	*Instantaneous*
m	*Motor*
M	*Mission, measure*
on	*Closed*
off	*Open*
p	*Poles*
P	*Supply, Active*
pp	*Peak to peak*
q	*Quadrature*
Q	*Reactive*
r	*Rotor, ring*
R	*Rectified*
ref	*Reference*
s	*Offset, stator, sun*
S	*Apparent*
sv	*Servovalve*
t	*Thermal*
w	*Winding*
+, -	*Positive, negative*

Acronyms

ACE: Actuator Control Electronics
ACS: Aircraft Combat Survivability
ADHF: Adaptive Dropped Hinge Flaps
AFCS: Automatic Flight Control System
AFDX: Avionics Full-DupleX Ethernet switching

AMP:	Arbitration on Message Priority
AP:	Autopilot
ARP:	Aerospace Recommended Practice
ART:	Actuator Remote Terminal
ASM:	Anonymous Subscriber Messaging
ATRU:	Auto Transformer Rectifier Unit
AWG:	American Wire Gauge
BCM:	Backup Control Module
BLDC:	BrushLess Direct Current
BPS:	Backup Power Supply
BSCU:	Braking and Steering/System Control Unit
BTV:	Brake To Vacate
BUCS:	Back-Up Control System
BYDU:	Backup Yaw Damper Unit
CAN:	Controller Area Network
CAS:	Control Augmentation System
CMM:	Control and Monitoring Module
COTS:	Commercially Off-The-Shelf
CPM:	Core Processing Module
CRV:	Crew Return Vehicle
CSMA:	Carrier Sense Multiple Access
DDV:	Direct Drive Valve
DFF:	Dynamic Force Feedback
DFS:	Differential Flap Setting
DPF:	Dynamic Pressure Feedback
DTSA:	Dynamic Time Slot Allocation
EAHA:	Electro-Assisted-Hydrostatic Actuator

EBA:	Electric Brake Actuator
EBAC:	Electric Brake Actuator Controller
EBHA:	Electro-Backup-Hydrostatic Actuator
EBPSU:	Electric Brake Power Supply Unit
EDP:	Engine Driven Pump
EHA:	Electro-Hydrostatic Actuator
EHA-FD:	Electro-Hydrostatic Actuator – Fixed Displacement
EHA-VD:	Electro-Hydrostatic Actuator – Variable Displacement
EHM:	Electro Hydrostatic Module
EHSV:	ElectroHydraulic ServoValve
EIS:	Entry into service
EMA:	Electro-Mechanical Actuator
EMF:	ElectroMotive Force
EMI:	ElectroMagnetic-Interferences
EMP:	ElectroMagnetic Pulse/Electro Mechanical Pump
ETRAC:	Electrical Thrust Reverser Actuation Controller
ETRAS:	Electric Thrust Reverse Actuation System
FbL:	Fly-by-Light
FbLW:	Fly-by-Less Wire
FbW:	Fly-by-Wire
FCC:	Flight Control Computer
FD:	Flight Director
FDIR:	Fault Detection Isolation and Reconfiguration
FOC:	Field-Oriented Control
GLA:	Gust Load Alleviation
HALT:	Highly Accelerated Life Testing
HCF:	Highest Common Factor

HIRF:	High Intensity Radiated Field
HSA:	Hydraulic Servo-Actuator
HUM:	Health and Usage Monitoring
HVDC:	High-Voltage Direct Current
IAP:	Integrated Actuator Package
IBC:	Individual Blade Control
IGV:	Inlet Guide Vane
IMA:	Integrated Modular Avionics
IPDU:	Integrated Power Distribution Unit
IXV:	Intermediate eXperimental Vehicle
LAF:	Load Alleviation Function
LCM:	Least Common Multiple
LEHGS:	Local Electro-Hydraulic Generation System
LGERS:	Landing Gear Extension–Retraction System
LRU:	Line Replaceable Unit
LVDS:	Low-Voltage Differential Signaling
LVDT:	Linear Variable Differential Transformer
MBD:	Model-Based Design
MCE:	Motor Control Electronics
MCU:	Motor Control Unit
MDEL:	Message DEscriptor List
MLA:	Maneuver Load Alleviation
MLCU:	Motor Lane Control Unit
MPD:	Motor Power Drive
MPE:	Motor Power Electronics
MTBF:	Mean Time Between Failure
OPV:	Optionally Piloted Vehicle

PbW:	Power-by-Wire
PCM:	Power Core Module
PDIV:	Partial Discharge Inception Voltage
PDU:	Power Drive Unit
PFC:	Power Factor Corrector
PFCU:	Powered Flying Control Unit
PLC:	Power line communication
PMSM:	Permanent Magnet Synchronous Machine
POD:	Power Over Data
PTLU:	Pedal Travel Limitation Unit
PWM:	Pulse-Width Modulation
RAE:	Remote Actuator Electronics
RAT:	Ram Air Turbine
RCCB:	Residual Current Circuit Breaker
RDC:	Remote Data Concentrator
REU:	Remote Electronic Unit
RMS:	Root Mean Square
RPA:	Remotely Piloted Aircraft
RTLU:	Rudder Travel Limitation Unit
RUL:	Remaining Useful Life
RVDT:	Rotary Variable Differential Transformer
SAE:	Society of Automotive Engineers
SAS:	Stability Augmentation System
SbL:	Signal-by-Light
SbW:	Signal-by-Wire
SbWL:	Signal-by-WireLess
SMCS:	Structural Mode Control System

SPS:	Service Propulsion System
SRU:	Smallest Replaceable Unit
SSAP:	Survivable Stabilator Actuator Package
STOF:	Start Of Frame
TDMA:	Time Division Multiple Access
THS:	Trim Horizontal Stabilizer
TRL:	Technology Readiness Level
TRPU:	Thrust Reverser Power Unit
TTP:	Time Triggered Protocol
TVC:	Thrust Vector Control
UAV:	Unmanned Aerial Vehicle
VC:	Variable Camber
VMC:	Vehicle Management Computer
VSCS:	Vertical Stabilizer Control System
WELS:	Wireless Emergency Lighting System

Index

A, B, C

anti-rotation, 172, 176, 192, 202
AS-5643, 49, 50
braking, 2, 35–37, 72, 73, 96, 114, 185–187, 190, 195, 213
brushless motor, 72, 98, 99, 195
CAN, 45, 46
charging, 163, 164
chopper, 99, 100, 103–105, 113, 114
cogging, 118, 126
COM/MON, 35
commutation cell, 104, 105
compliance, 36, 58, 73, 199, 215
concentrated, 107, 124–126, 128
conduction, 85, 87, 90, 97, 119, 121, 122, 133, 162
control augmentation system, 13, 14
control of power, 32, 63, 64, 69, 71, 73
cooling, 63, 88, 130, 131, 133–135, 161, 167, 171
Corona, 90, 91

D, E, F

daisy chain, 44, 45
damping, 34, 73, 114, 124, 155, 171, 179, 180, 188, 196, 204, 211

DC-link, 99, 105, 114
DC motor, 94, 98, 104, 105, 173
declutching, 143, 145, 159, 171, 190, 204, 207, 209–212
determinism, 43, 50
dimmer, 99
direct-drive, 188
direct drive valve, 15, 69
displacement control, 69, 70, 143–145
dissymmetry, 168
dog-teeth, 204, 205
double voltage, 90
dual input, 9, 16, 23
dynamic force feedback, 176
dynamic pressure feedback, 20
eddy current, 121, 122, 124, 125
electro-hydrostatic actuator, 32, 139
electro-mechanical actuator, 171
electromagnetic interferences, 31, 33, 130
extension-retraction, 96, 131, 153, 155, 157, 187–189, 191, 204, 210, 211
field oriented control, *see* vector control, 106, 109, 111, 112
filling, 72, 155, 157, 163, 164
fixed displacement, 72, 78, 79, 143, 145–147

flux weakening, 98, 110, 126, 176
fly-by-wire, 9, 18–22
force fighting, 34, 97, 145
force summing, 34, 163, 196, 207, 209
full duplex, 44, 50

G, H, I

gear drive, 81, 82, 198
half duplex, 44
helicopter, 5–8
hydraulic servo-actuator, 32
hydromechanical actuator, 17, 18
hysteresis, 22, 121, 122, 125, 126
individual blade control, 194
integrated, 41
integrated actuator package, 76
integration, 63, 130, 200
inverter, 103–105

J, L, M

jamming, 206, 208, 209
lamination, 97
landing gear, 35, 153, 185
launcher, 173
leakage, 68
less wire, 39
local electro-hydraulic generation, 71, 72
LVDT, 24–26, 176, 194
magnet, 95, 98, 124, 127, 165, 174
maintenance, 41, 171, 173, 187, 192, 195, 208, 209
military, 2, 3, 5, 22, 29, 48, 69, 70, 73, 76, 140, 148, 159
MIL-STD 1553B, 48
mode, 32, 35, 76, 80, 81, 101, 152, 153, 158, 185, 188, 190, 191, 215
multi-drop or multi-point, 45
mutualized, 30

N, O, P

nut-screw, 201, 207, 211
operation, 75, 96, 172
optical fibre, 39, 55, 56
overlapping, 125, 190
overload protection, 143, 171, 205, 215
park, 108, 109, 111
poles, 92–95, 106, 107, 116, 118, 124, 126, 128, 175, 195
position summing, 13, 15
power electronics, 20, 76, 84, 90, 95, 96, 98, 101, 112, 114, 116, 119, 121, 133
power factor, 114, 142
power management, 23, 73, 76, 203
power network, 63, 64, 72, 114
power view, 59
pump, 60, 61, 76, 78, 139, 155, 167–169
pump-up, 148, 164
PWM, 101, 117, 118, 190

R, S, T

rectifier, 113, 114
reducer, 198
redundant/redundancy, 128, 197, 211
remote, 30
ripple, 118
RS422, 46, 47
RS485, 46, 47
sensorless, 113
service life, 97
servovalve, 68
side-stick, 26, 27
signal view, 25, 40
simplex, 44, 155, 159
six-step, 105, 107, 116, 173
slots, 116, 124, 126
smart actuator, 8

stability augmentation system, 13, 14
steering, 188
switching, 119, 165
synchronization, 52, 145, 172, 209
temperature, 165
thermal balance, 119
thrust reverse, 195
top-down, 65
topology, 44, 46, 48, 50 52, 74

V, W

variable displacement, 139, 140
vector control, 106, 112, 125
wet, 122, 124, 167
wireless, 53, 58

Other titles from

in

Systems and Industrial Engineering – Robotics

2017

FEYEL Philippe
Robust Control, Optimization with Metaheuristics

RÉVEILLAC Jean-Michel
Modeling and Simulation of Logistics Flows 1: Theory and Fundamentals
Modeling and Simulation of Logistics Flows 2: Dashboards, Traffic Planning and Management
Modeling and Simulation of Logistics Flows: Discrete and Continuous Flows in 2D/3D

2016

ANDRÉ Michel, SAMARAS Zissis
Energy and Environment
(Research for Innovative Transports Set - Volume 1)

AUBRY Jean-François, BRINZEI Nicolae, MAZOUNI Mohammed-Habib
Systems Dependability Assessment: Benefits of Petri Net Models (Systems Dependability Assessment Set - Volume 1)

BLANQUART Corinne, CLAUSEN Uwe, JACOB Bernard
Towards Innovative Freight and Logistics (Research for Innovative Transports Set - Volume 2)

COHEN Simon, YANNIS George
Traffic Management (Research for Innovative Transports Set - Volume 3)

MARÉ Jean-Charles
Aerospace Actuators 1: Needs, Reliability and Hydraulic Power Solutions

REZG Nidhal, HAJEJ Zied, BOSCHIAN-CAMPANER Valerio
Production and Maintenance Optimization Problems: Logistic Constraints and Leasing Warranty Services

TORRENTI Jean-Michel, LA TORRE Francesca
Materials and Infrastructures 1 (Research for Innovative Transports Set - Volume 5A)
Materials and Infrastructures 2 (Research for Innovative Transports Set - Volume 5B)

WEBER Philippe, SIMON Christophe
Benefits of Bayesian Network Models
(Systems Dependability Assessment Set – Volume 2)

YANNIS George, COHEN Simon
Traffic Safety (Research for Innovative Transports Set - Volume 4)

2015

AUBRY Jean-François, BRINZEI Nicolae
Systems Dependability Assessment: Modeling with Graphs and Finite State Automata

BOULANGER Jean-Louis
CENELEC 50128 and IEC 62279 Standards

BRIFFAUT Jean-Pierre
E-Enabled Operations Management

MISSIKOFF Michele, CANDUCCI Massimo, MAIDEN Neil
Enterprise Innovation

2014

CHETTO Maryline
Real-time Systems Scheduling
Volume 1 – Fundamentals
Volume 2 – Focuses

DAVIM J. Paulo
Machinability of Advanced Materials

ESTAMPE Dominique
Supply Chain Performance and Evaluation Models

FAVRE Bernard
Introduction to Sustainable Transports

GAUTHIER Michaël, ANDREFF Nicolas, DOMBRE Etienne
Intracorporeal Robotics: From Milliscale to Nanoscale

MICOUIN Patrice
Model Based Systems Engineering: Fundamentals and Methods

MILLOT Patrick
Designing Human–Machine Cooperation Systems

NI Zhenjiang, PACORET Céline, BENOSMAN Ryad, REGNIER Stéphane
Haptic Feedback Teleoperation of Optical Tweezers

OUSTALOUP Alain
Diversity and Non-integer Differentiation for System Dynamics

REZG Nidhal, DELLAGI Sofien, KHATAD Abdelhakim
Joint Optimization of Maintenance and Production Policies

STEFANOIU Dan, BORNE Pierre, POPESCU Dumitru, FILIP Florin Gh., EL KAMEL Abdelkader
Optimization in Engineering Sciences: Metaheuristics, Stochastic Methods and Decision Support

2013

ALAZARD Daniel
Reverse Engineering in Control Design

ARIOUI Hichem, NEHAOUA Lamri
Driving Simulation

CHADLI Mohammed, COPPIER Hervé
Command-control for Real-time Systems

DAAFOUZ Jamal, TARBOURIECH Sophie, SIGALOTTI Mario
Hybrid Systems with Constraints

FEYEL Philippe
Loop-shaping Robust Control

FLAUS Jean-Marie
Risk Analysis: Socio-technical and Industrial Systems

FRIBOURG Laurent, SOULAT Romain
Control of Switching Systems by Invariance Analysis: Application to Power Electronics

GROSSARD Mathieu, REGNIER Stéphane, CHAILLET Nicolas
Flexible Robotics: Applications to Multiscale Manipulations

GRUNN Emmanuel, PHAM Anh Tuan
Modeling of Complex Systems: Application to Aeronautical Dynamics

HABIB Maki K., DAVIM J. Paulo
Interdisciplinary Mechatronics: Engineering Science and Research Development

HAMMADI Slim, KSOURI Mekki
Multimodal Transport Systems

JARBOUI Bassem, SIARRY Patrick, TEGHEM Jacques
Metaheuristics for Production Scheduling

KIRILLOV Oleg N., PELINOVSKY Dmitry E.
Nonlinear Physical Systems

LE Vu Tuan Hieu, STOICA Cristina, ALAMO Teodoro,
CAMACHO Eduardo F., DUMUR Didier
Zonotopes: From Guaranteed State-estimation to Control

MACHADO Carolina, DAVIM J. Paulo
Management and Engineering Innovation

MORANA Joëlle
Sustainable Supply Chain Management

SANDOU Guillaume
Metaheuristic Optimization for the Design of Automatic Control Laws

STOICAN Florin, OLARU Sorin
Set-theoretic Fault Detection in Multisensor Systems

2012

AÏT-KADI Daoud, CHOUINARD Marc, MARCOTTE Suzanne, RIOPEL Diane
Sustainable Reverse Logistics Network: Engineering and Management

BORNE Pierre, POPESCU Dumitru, FILIP Florin G., STEFANOIU Dan
Optimization in Engineering Sciences: Exact Methods

CHADLI Mohammed, BORNE Pierre
Multiple Models Approach in Automation: Takagi-Sugeno Fuzzy Systems

DAVIM J.Paulo
Lasers in Manufacturing

DECLERCK Philippe
Discrete Event Systems in Dioid Algebra and Conventional Algebra

DOUMIATI Moustapha, CHARARA Ali, VICTORINO Alessandro,
LECHNER Daniel
Vehicle Dynamics Estimation using Kalman Filtering: Experimental Validation

GUERRERO José A, LOZANO Rogelio
Flight Formation Control

HAMMADI Slim, KSOURI Mekki
Advanced Mobility and Transport Engineering

MAILLARD Pierre
Competitive Quality Strategies

MATTA Nada, VANDENBOOMGAERDE Yves, ARLAT Jean
Supervision and Safety of Complex Systems

POLER Raul *et al.*
Intelligent Non-hierarchical Manufacturing Networks

TROCCAZ Jocelyne
Medical Robotics

YALAOUI Alice, CHEHADE Hicham, YALAOUI Farouk, AMODEO Lionel
Optimization of Logistics

ZELM Martin *et al.*
Enterprise Interoperability –I-EASA12 Proceedings

2011

CANTOT Pascal, LUZEAUX Dominique
Simulation and Modeling of Systems of Systems

DAVIM J. Paulo
Mechatronics

DAVIM J. Paulo
Wood Machining

GROUS Ammar
Applied Metrology for Manufacturing Engineering

KOLSKI Christophe
Human–Computer Interactions in Transport

LUZEAUX Dominique, RUAULT Jean-René, WIPPLER Jean-Luc
Complex Systems and Systems of Systems Engineering

ZELM Martin, *et al.*
Enterprise Interoperability: IWEI2011 Proceedings

2010

BOTTA-GENOULAZ Valérie, CAMPAGNE Jean-Pierre, LLERENA Daniel, PELLEGRIN Claude
Supply Chain Performance / Collaboration, Alignement and Coordination

BOURLÈS Henri, GODFREY K.C. Kwan
Linear Systems

BOURRIÈRES Jean-Paul
Proceedings of CEISIE'09

CHAILLET Nicolas, REGNIER Stéphane
Microrobotics for Micromanipulation

DAVIM J. Paulo
Sustainable Manufacturing

GIORDANO Max, MATHIEU Luc, VILLENEUVE François
Product Life-Cycle Management / Geometric Variations

LOZANO Rogelio
Unmanned Aerial Vehicles / Embedded Control

LUZEAUX Dominique, RUAULT Jean-René
Systems of Systems

VILLENEUVE François, MATHIEU Luc
Geometric Tolerancing of Products

2009

DIAZ Michel
Petri Nets / Fundamental Models, Verification and Applications

OZEL Tugrul, DAVIM J. Paulo
Intelligent Machining

PITRAT Jacques
Artificial Beings

2008

ARTIGUES Christian, DEMASSEY Sophie, NERON Emmanuel
Resources–Constrained Project Scheduling

BILLAUT Jean-Charles, MOUKRIM Aziz, SANLAVILLE Eric
Flexibility and Robustness in Scheduling

DOCHAIN Denis
Bioprocess Control

LOPEZ Pierre, ROUBELLAT François
Production Scheduling

THIERRY Caroline, THOMAS André, BEL Gérard
Supply Chain Simulation and Management

2007

DE LARMINAT Philippe
Analysis and Control of Linear Systems

DOMBRE Etienne, KHALIL Wisama
Robot Manipulators

LAMNABHI Françoise *et al.*
Taming Heterogeneity and Complexity of Embedded Control

LIMNIOS Nikolaos
Fault Trees

2006

FRENCH COLLEGE OF METROLOGY
Metrology in Industry

NAJIM Kaddour
Control of Continuous Linear Systems

Printed and bound by CPI Group (UK) Ltd, Croydon, CR0 4YY
11/07/2022
03135556-0001